Cerebrum 2009

Cerebrum 2009

EMERGING IDEAS IN BRAIN SCIENCE

Dan Gordon, Editor

DANA
PRESS

New York • Washington, DC

Published by Dana Press, a Division of The Dana Foundation. Address correspondence to:
Dana Press
900 15th Street NW
Washington, DC 20005

DANA
PRESS

The Dana Foundation
745 Fifth Avenue, Suite 900
New York, NY 10151

DANA is a federally registered trademark.

Printed in the United States of America

10 9 8 7 6 5 4 3 2 1

ISBN-13: 978-1-932594-44-7
ISSN: 1524-6205

Art direction by Kenneth Krattenmaker.
Cover design and layout by William Stilwell.
Cover illustration by Mary Burr, courtesy of iStockPhoto.

www.dana.org

Contents

CEREBRUM 2009

Foreword

By Thomas R. Insel, M.D.

 **Thomas R. Insel, M.D.**, has served as director of the National Institute of Mental Health (NIMH) since 2002. His research in social neuroscience has used clues from comparative neurobiology to identify the molecular mechanisms, cellular substrates and neural systems involved in social attachments, from parental care to pair bonding. As NIMH director, Dr. Insel leads the nation's $1.4 billion investment in research on mental disorders with a mission of supporting discoveries that, according to the institute's strategic plan, "pave the way toward prevention, recovery and cure." Under Insel's leadership during what he has called the "decade of discovery," NIMH has been focused on applying insights from neuroscience and genetics research to counter the problems of people with serious mental disorders.

The brain is wider than the sky,
For, put them side by side,
The one the other will include
With ease, and you beside.

The brain is deeper than the sea,
For, hold them, blue to blue,
The one the other will absorb,
As sponges, buckets do.

The brain is just the weight of God,
For, lift them, pound for pound,
And they will differ, if they do,
As syllable from sound.

—EMILY DICKINSON, 1862

IMAGINE FOR A MOMENT that we are looking back from 2050 to consider the major scientific breakthroughs of the early 21st century. What will we see? The human genome, of course. Dark matter in the cosmos, perhaps. Renewable and sustainable energy sources, one hopes. But one of the most profound breakthroughs is so fundamental that we may take it for granted. Four centuries after Descartes, we are finally resolving the split between mind and brain, recognizing, as did Emily Dickinson, that the brain is "wider than the sky," wide and deep enough to contain the mind.

This has not been easy for us to accept. Just as the world looks flat, not round, the ineffable qualities of consciousness, imagination and memory all seem subjectively divorced from the three pounds of tissue in our heads. Yet Dickinson may have called it just right: "they will differ, if they do, / As syllable from sound." In 2050 this may well appear to be the singular insight of our era.

Modern neuroscience has put the brain and the mind back together, with far-reaching implications. Suddenly a range of interesting psychological problems, from dreams to moral decisions, becomes tractable via brain science. Rather than "reducing" psychology to biology, neuroscience research has consistently deepened and broadened psychological inquiry, helping us to understand aspects of mental experience that could not be discerned from subjective experience. Neurolinguistics, which maps the representations of grammar and distinct languages in the brain, is perhaps the most famous example of this phenomenon. Neuroimaging has not "reduced" our understanding of language; it has transformed it, revealing relationships neither intuitive nor accessible with psychological approaches. Cognitive neuroscience, social neuroscience, affective neuroscience and behavioral neuroscience have given us comparable insights, demonstrating that brain research is prone more to expanding frontiers than to reductionism.

But an even more profound implication of putting the brain and the mind back together is emerging in clinical research. For the first time, we can begin to address mental disorders as brain disorders. Although biological psychiatry has sought to do this for five decades, most of that effort has been mired in very simplistic models of brain function that considered, for example, depression as a "chemical imbalance" or schizophrenia as a "dopamine disorder." In the past decade, clinical neuroscientists have begun to study mental illnesses as disorders of brain circuits. This change may prove to be a fundamental transformation in how we study, diagnose and treat these very disabling illnesses. For instance, we may discover that the behavioral and cognitive manifestations of these illnesses are a late stage of a chronic brain process that could be detected years before psychosis or mood disturbance. Imagine the impact of predictive and preemptive medicine for schizophrenia or bipolar disorder. At the very least, we will need to begin training a new generation of mental health workers in cognitive and affective neuroscience so that they will be able to expand our understanding of mental disorders just as neuroscientists are expanding our understanding of mental life.

Cerebrum 2009 provides brilliant examples of how mental disorders can be addressed as brain disorders. David Spiegel explains how split personality or dissociative identity disorder arises from trauma. As he notes, "The problem is not that they have more than one personality, but rather that they have less than one—a fragmentation of self rather than a proliferation of selves." Spiegel describes this fragmentation as a form of severe, chronic, early-onset post-traumatic stress disorder. Recent neuroimaging research suggests that the traumatic experience may be out of sight but not out of mind. Understanding how the brain isolates traumatic memories is providing insights into how unconscious processes can have profound effects on observable behavior, including the fragmentation of the sense of self.

Remarkably, clinical neuroscience is helping us understand psychological treatments as well as psychological experience. As Elizabeth Norton Lasley reports in "A Road Paved by Reason," cognitive behavior therapy (CBT), first developed more than four decades ago by Aaron "Tim" Beck, has been shown repeatedly to be an effective treatment for mood and anxiety disorders. Lasley describes new research that looks at how CBT changes the brain as well as behavior. For its antidepressant effect, CBT appears to use different pathways than medication does, although there must be some final common targets for lifting depressed mood. By studying the neural effects of psychotherapy, we are getting our first glimpse of the pathways of recovery.

One common target for medication and CBT in the treatment of depression may be a little-known site deep in the brain's prefrontal cortex, Area 25. While rarely studied by neuroanatomists or neurophysiologists, Area 25 has emerged as a critical player in neuroimaging studies of depressed patients who recover, following either psychotherapy or pharmacologic treatment. As Jamie Talan notes in "Deep Brain Stimulation Offers Hope in Depression," neurologist Helen Mayberg has borrowed a treatment from Parkinson's disease to develop a new approach to treatment-refractory depression. For many (but not all) patients, deep brain stimulation leads to a lasting reduction in depressive symptoms. Welcome to the new age of addressing mental illnesses as brain disorders—beginning with recognition of the circuitry, identifying a hub for this circuit, manipulating this hub surgically and finally resolving the symptoms.

Elsewhere, one of the most perplexing areas where psychiatric and neurological symptoms overlap is traumatic brain injury (TBI). Although post-traumatic stress disorder (PTSD) has been a tragic result of recent wars and natural disasters, TBI is considered a relatively new problem as modern protective equipment allows more soldiers to survive severe head and blast injuries. TBI has been associated with cognitive and behavioral problems, often as disabling as PTSD, but there has been no diagnostic test or evident brain lesion identified. Wayne Gordon describes the consequences of unidentified TBI and the promise of a screening approach that should increase detection and awareness of this troubling and mysterious syndrome.

This edition of Cerebrum also includes reports on other important clinical problems for which neuroscience is yielding new, sometimes surprising insights. Richard J. Bonnie, Donna T. Chen, and Charles P. O'Brien describe the ethical dilemma presented by an effective new treatment for heroin addiction. Extended-release naltrexone may reduce relapse in addicts who have been incarcerated. Is it ethical to require parolees to take the drug? If this treatment is not a requirement, what is the most ethical means by which to implement it? Denise C. Park, addressing a different social concern, speaks to the challenge of maintaining neural health as we age. Does the adage "use it or lose it" apply to our brains? Park describes a theory of scaffolding in the aging brain that requires continuous activation to maintain brain health. How do we know which mental exercises are strengthening this scaffold and which efforts are wearing down an aging infrastructure? On a related topic, Guy McKhann and Brenda Patoine describe the common disabling problem of cognitive dysfunction following cardiac surgery. As they note, "What is good for the heart is good for the brain." With a longitudinal controlled study of patients undergoing cardiac surgery, McKhann and colleagues have demonstrated that some of the decline in cognitive function may be the result of neurovascular disease associated with coronary artery disease and not with the surgical procedure, as many believe.

While this edition of Cerebrum captures many of the frontiers where neuroscience is having an impact on human health, some of the most exciting areas of brain science are not yet ready for translation to the clinic. Note, for instance, Sebastian Seung's introduction to the emerging discipline of "connectomics," mapping the brain's wiring diagram. To understand how the brain functions we will need to know much more about how it is wired. Can we map the connections between all 100 billion neurons in the human brain? Not yet. But, as Seung explains, we can develop tools for this project in simpler brains, leading to a human connectome project that may someday reveal a comprehensive wiring diagram of our brains just as the human genome project provided a registry of our genes.

As neuroscience puts the brain and the mind together, we are seeing exciting new opportunities to make a difference for people with brain disorders. The need is great. These disorders, especially depression and addictive disorders, are the largest source of medical disability for young people in the developed world. With the aging of our population, dementia and stroke will become increasingly costly causes of medical morbidity and mortality. Brain science offers our best hope for reducing disability and ensuring a mentally healthy, lengthened life span. This edition of Cerebrum provides a superb introduction to the opportunities as well as the challenges that lie in translating neuroscience to the clinic. Enjoy.

CHAPTER 1

Coming Apart

Trauma and the Fragmentation of the Self

By David Spiegel, M.D.

David Spiegel, M.D., is the Willson Professor in the School of Medicine and associate chair of psychiatry and behavioral sciences at the Stanford University School of Medicine. He collaborated in the inclusion of Acute Stress Disorder, a new psychiatric diagnosis in the Diagnostic Statistical Manual, Fourth Edition (DSM-IV), and chaired the work group on Dissociative Disorders. Among his more than 400 scientific journal articles and chapters and 10 books, he is the editor of Dissociation: Culture, Mind and Body (American Psychiatric Press, 1994), and co-editor of Traumatic Dissociation (American Psychiatric Publishing, 2007). His research on the health effects of psychosocial support was the subject of a segment on Bill Moyers' Emmy Award–winning special Healing and the Mind. He is Past-President of the American College of Psychiatrists and the Society for Clinical and Experimental Hypnosis.

The controversial diagnosis of dissociative identity disorder (DID) has replaced what once was called "multiple personality disorder." People diagnosed with DID have trouble integrating their memories, sense of identity, and aspects of consciousness into a unified whole. New research supports the diagnosis and sheds light on what may have gone wrong in patients' brains, suggests David Spiegel, M.D. Spiegel, who chaired the professional working group that recommended the change of name in psychiatry's principal diagnostic manual, notes that the disorder likely stems from trauma and can be considered a severe form of post-traumatic stress disorder. Among the biological markers he describes are a smaller hippocampus and certain neurotransmitters. A better understanding of the importance of specific regions of the brain to memory and emotion may help push research forward.

IN POP CULTURE, "multiple personality disorder" is often portrayed as involving strategic, dramatic and seductive battles among personalities that are uncomfortably sharing one hapless body. On TV crime shows and in movies, the "split personality" is used as a dramatic excuse for mayhem or is feigned to evade criminal responsibility. Some believe that the disorder is the creation of credulous and overeager therapists. However, these and other common perceptions are mistaken. This article is written to set the record straight, to explain what this disorder is and what we understand about its causes, both in early life experience and in the brain. Some people do have what scientists now call "dissociative identity disorder" (DID), a name change made official in 1994, when the American Psychiatric Association published the fourth edition of its Diagnostic and Statistical Manual of Mental Disorders. Sufferers experience sudden loss of episodic memory; change from a sad, dependent, and helpless personality state to an angry, demanding, hostile one in seconds; and may find themselves in situations that they cannot understand. But they are the victims, not the authors, of their own fragmentation.

One "identity" may inflict physical damage on their body as "punishment" for another "personality" state, such as the patient who carved "I hate Mary," another of her identities, into her forearm with a knife. Mary was frightened and mystified about the injury. Such memory loss is often asymmetrical—one identity may be aware when another is prominent, but not vice versa.

The problem is not that there are "multiple personalities" existing in one body, as the old name of the disorder implied, but that the brain fails to integrate our different personae. We normally act like "different people" at work and at a party (hopefully), but we have continuity of memory and identity across the differences. Patients with DID do not. In fact, the problem is not that they have more than one personality but rather that they have less than one—a fragmentation of self rather than a proliferation of selves.

People with dissociative disorders are like actors trapped in a variety of roles. They have difficulty integrating their memories, their sense of identity and aspects of their consciousness into a continuous whole. They find many parts of their experience alien, as if belonging to someone else. They cannot remember or make sense of parts of their past.

Dissociative symptoms involving alterations in identity, memory, consciousness and body function are seen in cultures around the world, described as "ataques de nervios" in many Hispanic cultures and as states of trance and possession in China, Japan and India. DID is not all that rare. It affects some 1 percent of people in the United States, 0.5 percent in China, and 1.5 percent in Turkey and the Netherlands, according to various studies in these countries.

Controversy has swirled around the disorder, in part because it is extreme and dramatic. But new research has helped us understand the origins of this tragic condition, as well as how it is reflected in the brain.

Roots in Trauma

Evidence is accumulating that trauma, especially early in life, repeated and inflicted by relatives or caretakers, produces dissociative disorders. DID can be thought of as a chronic, severe form of post-traumatic stress disorder. The essence of traumatic stress is helplessness—a loss of control over one's body. The mental imprint of such frightening experiences sometimes takes the form of loss of control over parts of one's mind—identity, memory and consciousness—just as physical control is regained. During and in the immediate aftermath of acute trauma, such as an automobile accident or a physical assault, victims have reported being dazed, unaware of serious physical injury or experiencing the trauma as if they were in a dream. Many rape victims report floating above their body, feeling sorry for the person being assaulted below them. Sexually or physically abused children often report seeking comfort from imaginary playmates or imagined protectors, or by imagining themselves absorbed in the pattern of the wallpaper. Some continue to feel detached and disintegrated for weeks, months or years after trauma.

Abuse by a trusted authority figure such as a parent creates special problems. A child abused by a family member faces an ongoing dilemma: this beloved figure is inflicting harm, pain and humiliation, yet the child is both emotionally and physically dependent. The child has to maintain two diametrically opposing views of the same person, which creates considerable tension and confusion, a situation described by psychologist Jennifer Freyd as "betrayal trauma."[1] She showed that people prone to dissociation have selective amnesia for trauma-related words such as "incest." Freud wrote that "hysterics [his term for people prone to dissociation] suffer mainly from reminiscences." His point was that their often dramatic mental and physical symptoms were the product of early life trauma and conflict over sexually charged situations.

Can a Person Forget Trauma?

Humans process vast amounts of information. We can function only by being strategically selective in our awareness. To do otherwise would be like having every stored file in a computer open at once, or all the contents of one's office file cabinets spread out on the desk at the same time. Emotional arousal typically leads to increases in recall—most of us remember September 11, 2001, with more than average detail. However, we frequently try to control our emotional response to traumatic events, sometimes at the expense of recollection of them. Chelsea Clinton, who was living in Manhattan on 9/11, wrote in a magazine article that she started walking downtown toward the World Trade Center after the attack but hours later found herself uptown, with no memory of how she had gotten there.

Research bears out that blocking emotion about a trauma can also block memory of it. Neuroscientists Larry Cahill, James McGaugh and colleagues at the University of California–Irvine had volunteers watch slides of an accident. Before seeing the slides, one group was given a beta-blocker, a drug that blocks the stress-induced increase in heart rate and blood pressure triggered by the sympathetic nervous system. These subjects' arousal-related increase in recall was also blocked, compared with the recall of other subjects who were given a placebo rather than the beta-blocker.[2] Other research goes a step further, helping us understand what happens in the brain when we suppress memories. John and Susan Gabrieli and colleagues at Stanford and Michael Anderson at the University of Oregon[3] used positron emission tomography (PET), a sophisticated brain imaging technique, to study the brain's ability to inhibit memory. When participants were asked to block their memory of word associations, PET imaging showed increased activity in the dorsolateral portion of the prefrontal cortex, the part of the brain that enables us to stop and think, coupled with decreased activity in the hippocampus, the structure deep in the brain that controls memory storage and retrieval.

Evidence that this inhibition of memory happens in real life is more than anecdotal. Linda Meyer Williams[4] tracked down young women who had been treated in hospital emergency rooms for physical and sexual abuse an average of seven years earlier, during their childhood, and interviewed them about their history of trauma. Thirty-eight percent of them could not remember the episode that made a trip to the hospital necessary, although many discussed other episodes of abuse in detail. Another 14 percent reported that they had been unable to recall the traumatic episode for a period of time lasting months to years. One would think that anyone actually brought to a hospital emergency room for treatment would recall the necessitating episode, yet a substantial minority could not. While voluntary suppression of emotionally laden memories is less likely to be successful than suppression of neutral memories, psychologist Martin Conway of the University of Bristol in England has found that when people are motivated to forget, they are more likely to do so for trauma-related memories than for neutral ones.[5]

The pressure to forget is greater when children are abused by a trusted caregiver, who might cue memory retrieval unavoidably. The only way to prevent persistent recall of damaging memories would be to adapt internally and to deliberately avoid thinking of such memories—in Freud's terms, to push them away from consciousness. A study published in 2007 by Elke Geraerts and colleagues at Maastricht University in the Netherlands[6] provides additional evidence that some people simply do not persistently remember traumatic experiences. Forty percent of their sample of 98 people who responded to a newspaper advertisement about an abuse history reported discontinuous memories of it.

Why does this happen? For one thing, people naturally enter an unusual mental state during traumatic experiences. Their attention is narrowly focused. "The prospect of the gallows concentrates a man's mind wonderfully," Samuel Johnson famously noted. Mugging victims can often give a precise and detailed description of the assailant's gun but can describe little about his face. Dissociation can further isolate memories by separating them from common associative networks in the brain that would make associative memory retrieval easier. Thus trauma can elicit dissociation, complicating the necessary working through of traumatic memories. The nature of the acute response may influence long-term adjustment.

Often people who have suffered trauma consciously try to suppress their recollection of the painful events. Over time the forgetting becomes automatic rather than willful, in the same way that riding a bicycle requires a great deal of conscious mental and physical effort during the learning phase but becomes automatic over time.

Trauma can be conceptualized as a sudden discontinuity in experience: one minute everything is fine; the next, one is in serious danger. This may lead to a process of memory storage that is similarly discontinuous with the usual range of associated memories, which might explain the "off/on" quality of dissociative amnesia and its reversibility with techniques such as hypnosis. Though dissociated information is out of sight, however, it is not out of mind. The information kept out of consciousness nonetheless has effects on it.

Insight from Post-Traumatic Stress Disorder

Many people suffering from PTSD are unable to recall important aspects of the trauma. Others feel detached or estranged from people afterward. Emory University psychiatrist Douglas Bremner found high levels of dissociative symptoms among Vietnam veterans with PTSD, and they also reported dissociating during combat.[7]

In a sample of 122 women seeking treatment for childhood sexual abuse, my research team found that a majority (66, or 54 percent) experienced PTSD symptoms. These women had more dissociative symptoms than those who did not evidence PTSD symptoms.[8] Furthermore, among those with PTSD, dissociative symptoms were associated with higher levels of childhood abuse. Those with symptoms of dissociation also had more symptoms of physiological hyperarousal, such as a pronounced startle response after hearing a loud noise, suggesting that there is an association between psychological avoidance and physiological hyper-reactivity.

However, other studies provide evidence that dissociative detachment after a traumatic experience numbs the body as well as the mind. Psychologists Michael Griffin, Patricia Resick and Mindy Mechanic at the University of Missouri studied women who had been raped. Within two weeks of the rape, women with PTSD resulting from the assault who reported high levels of dissociation during the rape had smaller increases in heart rate and skin conductance (each a measure of the autonomic nervous system's stress response) during exposure to trauma-related memories. The women with PTSD but lower levels of dissociation responded with larger increases.[9] Similarly, neuroscientist Ruth Lanius at the University of Western Ontario in Canada[10, 11] studied people with PTSD and dissociative symptoms resulting from sexual abuse. Those with high levels of dissociation showed no increase in heart rate when read scripts with vivid descriptions of their trauma but had activation in the prefrontal cortex (which is responsible for thought and inhibition) and parts of the limbic system (which is responsible for emotion) on functional magnetic resonance imaging scans. Those with lower levels of dissociation responded with increased heart rates and less activity in those brain regions during this task.

Other studies reveal a distinction between the body's immediate, neural stress response and the secondary, hormonal response. Dissociation after trauma is linked with higher levels of cortisol (a stress hormone that mobilizes glucose into the blood to assist with the fight-or-flight response) in the saliva, according to research in which cortisol levels were measured 24 hours after a stressful interview among adult women who were sexually abused during childhood.[12] So while the immediate neural stress response system is suppressed by dissociation, the secondary hormonal stress response system is triggered by it.

What Happens in the Brain

Dissociative disorders involving fragmentation of identity, memory and consciousness seem less mysterious if we conceptualize identity as the product of mental effort rather than a given—a bottom-up rather than a top-down model of how the brain processes information. Neural systems that process the coincident firing of millions of neurons at a time must extract coherence from all this activity, and it is not surprising that in some cases these systems do not succeed. Neurons that fire together wire together, but building large, complex and yet coherent neural networks may not always lead to a coherent sense of identity. Factors that restrict neurons from firing in association may limit the continuity of identity that emerges from experience and memory.

Hippocampal Volume

Another plausible neurobiological mechanism linking childhood trauma to dissociative difficulties with the integration of memory is smaller hippocampal volume. As mentioned earlier, the hippocampus, part of the limbic system situated in the middle portion of the temporal lobe, organizes memory storage and retrieval. The hippocampus is rich in glucocorticoid receptors, which are sensitive to stress-induced exposure to cortisol. Researchers have provided strong evidence in animals that early life experiences have lasting effects on the hormonal stress response system, either making it unduly sensitive to stress or protecting it from overreaction throughout life. Studies in humans show that, while minor stressors may produce resilience, childhood sexual abuse does the opposite: it sensitizes the individual to subsequent stressors decades later. This research indicates that chronically elevated cortisol levels may damage the hippocampus, leading to smaller size and poorer function.

Imaging studies by Murray Stein at the University of California, San Diego, and Eric Vermetten at Utrecht University in the Netherlands have shown that people with a history of childhood abuse and dissociative disorders indeed have smaller hippocampi, and that the reduction in size correlates with the severity of dissociative symptoms.[13, 14] Vermetten also found reductions in the size of the amygdala, the seat of fear and anger conditioning. Researcher Douglas Bremner found similarly smaller hippocampal size among veterans with PTSD symptoms. However, Harvard psychiatrist Roger Pitman proposed an alternative explanation for this relationship.[15] He studied 35 pairs of identical twins, one of whom had been exposed to trauma and one of whom had not. Pitman found that smaller hippocampal volume is indeed a risk factor for PTSD severity, but is not affected by exposure to trauma. A smaller hippocampus, he reasoned, may underlie vulnerability to the development of PTSD symptoms rather than occurring as a result of trauma exposure.

In any case, a smaller hippocampus would likely limit a person's ability to encode, store and retrieve memories and manage the emotions associated with them. The hippocampus is a context generator, helping us to put information into perspective. Wolf has shown that activity in the hippocampus buffers the effects of stressful input on the hormonal stress response system.[16] Ruth Lanius demonstrated that those who dissociate in response to listening to accounts of their traumatic experiences have decreased activity in the brain adjacent to the hippocampus—they remember less and their brain memory systems are less active.[11] Limitations on hippocampal size and function hinder memory processing and the ability to comprehend context, especially in light of contradictory memory encoding and storage. Among patients with PTSD and dissociative symptoms, research also indicates that there is higher connectivity between two portions of the brain—the right insula and the left ventrolateral thalamus—that are involved in perception of bodily processes and emotion and consciousness. This finding provides further evidence that mental distress and physical distress are both triggered by traumatic memories.

Neurotransmitter Activity

Neurotransmitters convey information from one nerve cell to another, and a specific one may be involved in dissociation. It has long been known that drugs that block the activity of the N-methyl-d-aspartate (NMDA) subtype of glutamate receptors in cortical and limbic brain regions produce dissociative symptoms, perhaps via a one-time release of glutamate. Anti-anxiety medications such as lorazepam stimulate the release of gamma amino butyric acid (GABA), a neurotransmitter that inhibits rather than stimulates activity in many regions of the brain. Yale researcher John Krystal has suggested that GABA may also play a role in dissociative symptoms. His work suggests that administering a drug that stimulates GABA increases dissociation.[17, 18]

Coming Together:
Future Research on Dissociation

Two heads are not better than one when they share the same brain. The fragmentation of mental function that can occur after a series of traumatic experiences may both protect a person from distress and make it harder for the individual to put the trauma into perspective. As we come to appreciate the complexity of neural development, we also understand that early life experiences have a profound effect on the developing brain. In dissociation, achieving a sense of mental unity is such a difficult task that it can be disrupted by events that challenge body integrity, emotional control and the development of relationships. Future research will reveal more about specific genetic vulnerabilities that may make certain individuals especially susceptible to the disorganizing effects of traumatic stress.

We also need to understand more about neural development and function: how do specific regions of the brain facilitate or inhibit memory, emotion and their interaction? How can we use this knowledge to better treat individuals suffering from dissociation? Current treatments primarily involve psychotherapy, and increasing knowledge of brain structure and function may provide necessary connections for therapists and their patients, helping patients to understand and control their dissociative tendencies while working through the consequences of traumatic experiences. Other research may lead us to a specific medication that treats uncontrolled dissociation; at present there is none.

As we better understand control systems in the brain that underlie dissociation, we hope to enable people so that their response to trauma does not reinforce feelings of helplessness but rather augments their control over their identity, memory and consciousness.

CHAPTER 2

Your Brain and Heart Surgery

By Guy McKhann, M.D., and Brenda Patoine

Guy McKhann, M.D., is professor of neurology and neuroscience at the Zanvyl Krieger Mind/Brain Institute, Johns Hopkins University, Baltimore. He is the founding chairman of the Department of Neurology at Johns Hopkins. He also serves as scientific consultant for the Dana Foundation.

Brenda Patoine is a freelance science writer who has been covering neuroscience for more than 15 years. She writes regularly for Annals of Neurology and The NCRR Reporter, a publication of the National Institutes of Health. On the Web, her work can be found on AARP.org (Staying Sharp series), and on alzforum.org and alzinfo.org, two Alzheimer's Web sites.

Some heart surgery patients have cognitive problems afterward, but the idea that the surgery causes these problems is misleading. Guy McKhann and Brenda Patoine reveal how brain health before surgery appears to be the key influence on brain health afterward; recent findings suggest that pre-surgery problems affecting the arteries of the heart also affect the blood vessels feeding the brain. These findings have important implications for treatment options and how doctors inform patients of risks. Last but not least, they reinforce the idea that what's good for the heart is good for the brain, too.

SINCE THE EARLIEST DAYS OF CARDIAC SURGERY— unquestionably one of the great medical triumphs of the late 20th century—post-surgery memory problems and mental fogginess have been observed in patients. Heart doctors even had a nickname for it: "pump head." The slang referred to the heart-lung bypass machine that made the surgeries possible but was presumed to be the cause of the problem.

Fortunately, such short-term cognitive problems usually resolve within a few weeks; by a year after surgery, most people are back to normal. It is also not at all clear that the "pump" itself is the culprit, since cognitive problems are also associated with newer "off-pump" procedures.

But the biggest controversy surrounding heart disease and the brain is the question of longer-term cognitive problems. It has become conventional wisdom that, after bypass surgery, some people have cognitive dysfunction. In practice, this message often gets oversimplified, leaving the impression that bypass surgery in fact causes cognitive decline. Indeed, the perceived cognitive risks of this surgery influence treatment decisions all the time.

As it turns out, this line of thinking not only oversimplifies a complex problem, it also may in fact be downright misleading.

My research group's most recent data suggest something quite different. We found that over the long term, people with documented coronary artery disease have essentially the same rate of cognitive problems whether they have bypass surgery or some less invasive procedure. Although people who have cardiac surgery to remedy their heart disease do generally experience modest cognitive decline over time, the rate of decline is not significantly different for them than for people whose heart disease is treated with stents, for example. In fact, the best predictor of whether a person will suffer latent cognitive decline is not whether they have bypass surgery but how healthy their brain was before any intervention.

Taken together, these findings suggest to us that cognitive decline occurs in these patients—all of whom clearly have heart disease—because the same disease processes that have gradually closed their heart arteries have also affected blood vessels feeding their brains. Our working hypothesis is that surgical patients who go on to suffer cognitively started out with some level of cerebrovascular disease to begin with—it just may not have been far enough along to produce noticeable clinical symptoms. Several lines of evidence support this hypothesis.

The implications of this concept, should it prove true, are important. We may need to carefully rethink how we screen people for surgery, how we talk to people about the risks of surgery and what needs to be done to lower risk factors that are modifiable. In addition, this hypothesis would affect the contentious debate raging right now about the comparative risks and benefits of bypass surgery versus nonsurgical interventions such as stenting, in which a catheter embedded with a mechanical device is threaded through an artery in the groin to fix problem arteries in the heart. The risk of latent cognitive decline with surgery has entered into this debate and is often cited by proponents of nonsurgical techniques as one more reason their approach is better.

There are public health implications as well. Despite decades of health messages urging otherwise, Americans are notorious for lifestyles that increase the risk of heart disease, such as eating too much fat and not getting enough exercise. Perhaps we will pay more attention if we understand that a heart-healthy lifestyle also might help save our brains.

I don't know of anyone who desires to spend their golden years in the fog of cognitive dysfunction or dementia, and baby boomers seem to be particularly inclined to take steps to protect their mental health. Such protection could go a long way toward improving public health and controlling health-care costs, given the dire predictions about the looming impact of age-related brain diseases as boomers grow old. The message is clear: do your heart good and you do good for your brain as well.

Five Decades of Fixing the Heart

When it was introduced in the 1950s, the cardiopulmonary bypass machine, or "heart-lung machine," changed cardiology forever. Pioneered by John Gibbon in collaboration with his wife, Mary, and engineers from IBM, it was first used successfully in 1953 on 18-year-old Cecilia Bavolek to close a hole between the upper chambers of her heart. Refinements continued rapidly as surgeons embraced the technique, and the era of cardiovascular surgery was born.

The bypass machine acts as a temporary stand-in for both heart and lungs, pumping blood and oxygen through the patient's body even while the heart is stopped. This makes it possible to operate directly on heart valves, chambers or the aorta, the major vessel that routes blood out of the heart. Temporarily stopping the heart from beating also stills it so delicate surgical procedures can be performed more precisely.

In bypass surgery—technically called coronary artery bypass grafting (CABG)—the surgeon literally sews in a grafted vein, usually from the patient's leg, to bypass blockage in one or more of the arteries that supply the heart. The evidence is clear that CABG can be remarkably effective in relieving chest pain (angina), a common symptom among people with heart vessel diseases.

Until a few years ago, CABG surgery was one of the most common procedures performed in many hospitals in the United States; it was not unusual for major medical centers to do 1,000 to 2,000 a year. In its heyday, roughly 800,000 CABG surgeries were done annually, but that number is now down to about 250,000, largely because of the increased use of stents.

Stents are small tubes that are placed inside blood vessels to keep them open. Modern stents are often impregnated with drugs or chemicals designed to keep them functioning longer once they are implanted. While stents have not yet displaced CABG, they have become the treatment of choice for many physicians when coronary artery disease is limited to one or two blood vessels.

Some surgeons have also turned to off-pump surgery, in which the bypass pump is not used and surgery is done on the beating heart as an alternative to CABG. They hoped that this approach would have fewer brain complications, but the results have been mixed: off-pump surgery is associated with fewer strokes but does not appear to offer much benefit in terms of reducing longer-term cognitive deficits.

Paradoxically, as a result of the rise of stents and off-pump surgery, the people who are treated with CABG today tend to be the same ones who are at greater risk for latent cognitive problems: they are typically older and have more risk factors for vascular disease.

My colleagues at Johns Hopkins University Medical School and I first became interested more than 15 years ago in finding out what happens to the brains of people who have had CABG procedures. In our experience, cognitive difficulties arise shortly after surgery in about 30 percent of patients. The underlying biological mechanisms are not entirely clear, but many factors are likely to be involved. Older people and those who have symptoms of brain swelling ("encephalopathy") as a result of surgery are more likely to have short-term problems, and surgery patients should expect their doctors to discuss the possibility of short-term changes in cognition with them, as well as with their family members and friends. The good news is that people get better, many within three months of surgery and most within a year.

Thus we have turned our attention to problems over the longer term.

Rethinking Latent Decline

During the past several years, considerable controversy has arisen over whether a person who undergoes CABG is more likely in succeeding years to have progressive decline in cognitive function or heightened risk for Alzheimer's disease or other types of dementia. Several studies have reported long-term cognitive changes after CABG, including an oft-cited

2001 report in the New England Journal of Medicine by Mark Newman and colleagues, who found significant cognitive decline in more than 40 percent of CABG patients five years post-surgery.

This and other studies have fueled a widespread belief in the "CABG-equals-cognitive-decline" mantra. The problem is that the studies on which this belief is based did not include any control groups with which surgical patients were compared. For us, this brings up the question "Cognitive decline compared with whom?"

My colleagues and I have recently completed a long-term prospective study in which we have tracked for six years people who had coronary bypass grafting. We compared them with people who have proven coronary artery disease but did not have surgery (a group we call "nonsurgical controls," most of whom had stent procedures) and with a group of people we call "heart-healthy controls" because they have no known risk factors for coronary artery disease.

We performed a battery of cognitive tests on everyone at the study's start, before any interventions were performed, and then at time points of one, three and six years afterward. The "baseline" analysis at the study's start proved very interesting in and of itself.

It turns out that the people in our study who had existing coronary artery disease, whether or not they went on to have surgery, were quite different as a group at baseline than the heart-healthy controls. They performed more poorly on various tests of cognitive function, particularly on tests that gauge executive function, which encompasses abilities related to planning, making decisions and controlling impulses. Memory function, in contrast, was quite similar among the groups at the beginning of the study.

After one year, all three groups showed some improvement in the cognitive testing, most likely because of a practice effect (i.e., they got better at taking the test each time they took it). But the two groups with coronary artery disease didn't improve as much as those in the healthy control group. In other words, their cognitive scores were comparable with one another but not on a par with those of the heart-healthy controls. At six years, we do see a moderate decline in cognitive function, but it is essentially the same in both groups with coronary artery disease. There is no long-term decline specific to those who had surgery. A logical explanation for this is that blood vessels throughout the body and brain, not only the heart, had already been damaged by arterial disease. It was this ongoing disease process—not the surgery per se—that set these people up for later cognitive decline.

Predicting Stroke and Other Risks

Although CABG surgery seems not to be the cause of cognitive decline in cardiac patients, extra care may be necessary in patients who are at risk for stroke and other problems. Strokes occur when an area of the brain does not get enough blood. Blood supplies oxygen, and the brain needs oxygen to function. Strokes can be caused by either a hemorrhage in the brain or the blockage of a brain blood vessel. Think of the analogy of a sprinkler system on a lawn: if one of the pipes in the system becomes clogged, that area of the lawn does not get water and eventually dries up and dies. A similar process occurs in the brain if a blood vessel becomes clogged and blood flow to a region of the brain is restricted.

Why are strokes associated with CABG and other forms of heart surgery? The most likely explanation is that during surgery, small bits of material, usually blood clots or bits of tissue associated with arteriosclerosis (so-called "hardening of the arteries"), break off and are carried through the bloodstream to the brain. There, they lodge inside blood vessels and eventually block blood flow to the brain. Stroke can also occur when blood pressure falls too low to pump enough blood to the brain. For these reasons, blood pressure is carefully monitored during surgery. In most people who suffer a stroke as a result of heart surgery, bits of material and low blood pressure seem to be working together to cause damage.

By studying the characteristics of people who have had a stroke associated with CABG, it has been possible to develop a paradigm to estimate an individual's chance of having a stroke based on identified risk factors. Age is one important factor. A history of hypertension or diabetes is another because these conditions are associated with vascular disease. A previous stroke or narrowing of blood vessels in the neck or the legs also indicates heightened risk. For those at higher risk for stroke, we advocate a preoperative magnetic resonance imaging study. However, the information we would most like to have, the status of the small blood vessels of the brain, is not obtainable by current imaging methods.

Also sometimes associated with cardiac surgery is encephalopathy, a pathological process involving large areas of the brain. People with encephalopathy are generally slow to wake up from the anesthesia used during surgery and are often quite confused—even delirious—when they do wake up. They may be combative and have problems with memory, in some cases not even recognizing family members. Postoperative management for these patients can be challenging.

We consider stroke and encephalopathy to be part of a continuum with common underlying mechanisms—namely, lack of blood flow to the brain because of low pressure ("hypo-perfusion") and/or multiple small strokes. The same kind of predictive modeling that is used for stroke can be used to estimate an individual's risk for encephalopathy. Both stroke and encephalopathy are associated with poor recovery and increased risk of death.

Reducing Surgical Risks to the Brain

In a perfect world, all surgical candidates would undergo testing that would determine the extent of damage and disease within their neurovascular systems. In that world, we would also have medications that could be used prior to surgery to protect the brain from injury. Preventing injuries would be very difficult, but changing the brain's reaction to the injury—so-called "neuroprotection"—is a realistic goal.

There have been many attempts to develop neuroprotective drugs, but despite promising results in experimental animals, very little has been transferred to the human. Large-scale clinical trials have failed, without exception. But cardiac surgery patients are an excellent group in which to evaluate neuroprotective drugs because we can take baseline measurements before surgery, carefully control what happens during surgery and track patients over time to determine if there is a protective effect and how it might be maximized. That is clearly a next step in research related to this surgery.

If a person is deemed to be at high risk for stroke or encephalopathy, several steps should be taken. The first is to reevaluate how badly he or she needs the surgery and to consider all other alternatives, such as unblocking the artery with a stent. The second is to alter the surgical procedure, which might involve placing filters in the bypass system to prevent clots from reaching the brain. Alternatively, some surgeons have attempted off-pump surgery. However, the evidence that this approach is markedly better than conventional CABG is not conclusive.

Finally, people at high risk for complications require careful evaluation prior to surgery to determine, as best we can, the status of the blood vessels that supply the brain and those within the brain itself. New techniques for evaluating what we call brain perfusion—how much blood is being pumped to various parts of the brain—hold promise for providing necessary information about the status of someone's brain before or after surgery.

In the meantime, there are things that can be done to reduce the risk of cognitive complications, not just from surgery but in general. And guess what? They are the same things we should be doing to improve the blood vessels of the heart. Perhaps you've heard the line: what's good for the heart is good for the brain. This is a fundamental tenet that modern neuroscience has taught us. Stopping smoking, increasing our level of physical exercise and keeping our weight under control are essential components of a "brain-healthy" lifestyle, as is proper management of risk factors such as hypertension, diabetes and high cholesterol. This sounds obvious, but many patients who have surgery go back to the very lifestyle that got them into trouble in the first place.

CABG is clearly a very valuable procedure and an important treatment option for people with serious heart disease. We now know that it is possible to predict who is likely to sail through the procedure without much trouble and who is at risk for having cognitive complications in the aftermath of surgery or much later. Anyone in one of these now well-defined high-risk categories who is facing heart surgery should have a serious discussion with his or her cardiologist and surgeon to explore possible alternatives and determine how to lower the risk as much as possible.

CHAPTER 3

Deep Brain Stimulation Offers Hope in Depression

By Jamie Talan

Science writer **Jamie Talan** (with co-author Richard Firstman) won the 1998 Edgar Award for best nonfiction for "The Death of Innocents," a gripping account of forensic science that was also a New York Times Notable Book of the Year. Talan, who covered neuroscience for Newsday for more than twenty years, is science-writer-in-residence at the Feinstein Institute for Medical Research in New York.

There is a new hope for patients who have severe depression. An experimental surgical procedure, deep brain stimulation, is proving to reverse the effects of unrelenting depression by stimulating a precise network of brain cells. Jamie Talan reveals how some of the top scientists are using this procedure.

DEANNA COLE-BENJAMIN'S DEPRESSION sneaked into her world through the back door of her happy, balanced life. By her mid-30s, she had a job as a public health nurse, a husband and three children. Growing up in Ottawa, Ontario, she knew nothing about deep sadness. But as the new millennium began, depression descended without warning. Years of traditional therapy were of little help; finally, an experimental surgical procedure to implant stimulating electrodes into the white fibers in her brain made it possible for her to find a way back.

If depression had to shatter her world and push her into a psychiatric hospital for four years off and on, it arrived at a propitious moment in the history of modern medicine. Deep brain stimulation, or DBS, for depression was a technique borrowed from the world of movement disorders that showed hints of working for psychiatric conditions in some patients but no proof—yet. Deanna took her place on the operating table only a year into the first attempts to stop unrelenting depression by stimulating a precise network of brain cells. And it worked.

The practice of putting electrodes into the brain and electrically stimulating it at high frequency to calm abnormal hyperactive networks has helped patients with Parkinson's disease, essential tremor and dystonia for more than 15 years. But using the technology to treat depression developed from the pioneering work of Helen Mayberg, a neurologist who began her career when brain scanning technology promised to reveal the secrets of the sick brain.

First, Mayberg mapped the depressed brain on medications, then on therapy and then on a placebo pill. Each step of the way, she carefully charted the brain as if it were a city of streets and avenues. She realized that treatments took different roads but ultimately arrived at the same address. That was why people with depression could get better many different ways, even with a placebo pill.

The limbic structures that regulate mood feed into the frontal cortex, striatum, thalamus, hypothalamus and brain stem. These regions communicate with one another all the time, and problems in the circuit can lead to difficulty with thinking, attention, mood and behavior. Mayberg found that these circuits, particularly a hyperactive network of brain cells in the subgenual cingulate region, also called Brodmann area 25, are abnormally overactive in depression. When treatments work, the activity of these networks appears to return to normal. It made sense that so many brain areas are involved in depression, which encompasses more than just negative mood. People lose their motivation to get out of bed, to work, to love. Many have problems paying attention and thinking clearly. Eating and sleeping patterns can be way out of kilter.

For Mayberg, deep brain stimulation was the next logical step in figuring out how the network was broken. In the late 1990s, she approached Andres Lozano of the University of Toronto, who was well known for his work in movement disorders. Lozano had begun implanting stimulating electrodes in patients with Parkinson's and dystonia in the early 1990s. For many, the crippling symptoms of tremors, rigidity and slowed movements disappeared with a flick of the switch. Mayberg wondered if deep brain stimulation could alter mood and behavior as well.

"These are the circuits for depression," she told Lozano, pulling out a scan of the depressed brain. "Can we do something about it?"

By the time Deanna Cole-Benjamin came to the University of Toronto in 2004, Mayberg had already spent five years in Canada. For part of that time she worked with Lozano and psychiatrist Sidney Kennedy to design a small clinical study to test deep brain stimulation for patients who had exhausted all other treatment options. In 2002 they were granted approval to operate on five patients. Within a year they brought their first depressed patient into the operating room.

During the two-step surgical procedure, neurosurgeons use

computerized maps and brain scans to identify the precise target within the depression network. The patient's head is kept perfectly still in a metal frame, the skull bolted into the gear with screws to allow scientists a steady route into the brain. Surgeons use local anesthesia: the patient feels no pain, yet is awake during surgery so that the team can ask questions about his or her mood, thoughts or symptoms as they approach the target and can also keep a sharp eye on any potential side effects. Surgeons enter the brain through two holes in the skull and thread an electrode with four leads in the white matter tracts of the subgenual cingulate region on both sides of the brain. Once the electrode is in place, the surgeons send electricity to it to test its effects—the goal is to stimulate the network that is affected through that specific tissue target. After testing is completed, the patient is placed under general anesthesia, and a small battery pack is implanted near the collarbone and connected to the implanted leads and extension wires in the brain.

The system has an external controller so that doctors can program the device. The DBS team generally gives the patient a week or two to recover from the procedure before they bring him or her back for a programming session to begin chronic stimulation. Using the optimal contact established by a combination of the postoperative magnetic resonance images showing the location of the electrodes and the optimal effects seen during the electrode testing, the team sets standard starting parameters at that contact on each side of the brain. The patient returns weekly and rates the change in depression symptoms; then adjustments are made on the basis of the previous week's reported improvement. Generally, only minor adjustments in current are necessary and, once a consistent pattern of improvement emerges, the settings are kept stable.

Deanna was a strong candidate for the experiment: the depression exhausted her, and she felt indifferent to her family's love. She quit her nursing job in 2000, lost interest in the everyday activities of her life and spent much of her time in bed. "There was a mountain between me and everyone around me. I didn't have the strength to climb that mountain and reach them," she explained. That detachment and the exhaustion kept her under the covers that winter. As a nurse, she recognized that she was depressed but thought she should be able to shake it off. She couldn't. Two days before Christmas, her family doctor was so afraid that she might hurt herself that her husband drove her to Kingston Psychiatric Hospital, a large collection of stone buildings with bars on the windows and locks on the doors. It became her home, off and on, for nearly four years.

At Kingston, Deanna fantasized about swallowing enough pills to die or drowning in the chilly waters of Lake Ontario, which was the lone bit of beauty visible to patients. She spoke openly about her ideas; twice she grabbed a handful of pills and swallowed them in hopes that a permanent sleep would end her pain.

Doctors at the hospital tried every medicine and treatment that held some promise for alleviating depression. Over her four years in and out of Kingston she had 80 rounds of electroshock therapy and even a relatively new treatment at the time called transcranial magnetic stimulation, in which a magnetic current is delivered through the scalp to stimulate areas of the cortex at the surface of the brain. Nothing worked. "It was worse than being dead in a way," said Deanna.

By April 2004 she was out of the hospital and seeing Gebrehiwot Abraham every few days. Abraham, a psychiatrist at the hospital, had developed a close relationship with Deanna since she was first admitted in 2000. When she arrived at his office for one of the appointments, only weeks after her discharge, she knew that he would probably have to readmit her. The suicidal thoughts were back. But this time he handed her a fax that he had just received from Sidney Kennedy of the University of Toronto. Kennedy was recruiting patients with treatment-resistant depression for an experimental study on the deep brain stimulation procedure. He would find his match in Deanna Cole-Benjamin.

Upon meeting the former nurse, Kennedy agreed that she fit the very rigorous criteria set for the pioneering experimental treatment. She spent days taking a battery of neuropsychiatric tests and talking to many psychiatrists, psychologists and nurses. Two different brain scans provided insight into the structure and function of her brain under depression.

When Mayberg, who was working with Kennedy, met Deanna, she asked her what she wanted from the surgery. "To feel connected to my kids again," she said. "To feel their hugs and kisses." Mayberg agreed she was an ideal candidate for the study. Deanna and her husband, Gary, knew they had no other option but to embark on an experiment that had been tried in only a few other patients. She wondered whether she would turn into somebody different as a result of the procedure. And what if it didn't work?

The surgery was performed on June 7, 2004. In the operating room, doctors positioned the metal stereotactic frame, like a medieval halo, over her head and used screws to pin the frame into her skull, painlessly stabilizing it. A computer had already mapped out a trajectory to their target, the bundle of white matter fibers deep inside the brain near Brodmann area 25. Other than the local anesthesia used to prevent skin and skull pain, Deanna was not sedated, so that the medical team could observe any changes in behavior—good or bad—that might occur with testing of the contacts once the electrodes had been implanted.

The operating room was filled with the surgical and research team members. With Mayberg standing next to Deanna, midway between her head and feet, Lozano drilled a hole on both sides of Deanna's skull and made his way to the target. Deanna recalled the sound of the saw, the smell of burning bone, the static from a machine recording the activity of her brain. Lozano threaded the first electrode into the left part of her brain, the computer helping to mark his course. Then he headed for the right side of her brain, placing a second electrode in the mirror-image location. Deanna was withdrawn, quiet. Mayberg was asking her questions: "How do you feel? What are you thinking?" "Nothing," she replied—to both. Lozano turned on the stimulator, with a nod in Mayberg's direction.

Deanna recalled turning toward Mayberg, who was wearing blue surgical scrubs. The thought occurred to her immediately that she had been living in a black-and-white world. Now there was color. She looked into Mayberg's green eyes. They were warm. "I really feel like I know you," she told the neurologist.

Then Lozano turned the stimulating electrodes off and her world darkened again. She grew quiet. "It's gone," she said. "It must have been a dream."

They turned the stimulators on again. "Wow," she said. The room had brightened.

"We knew we were on to something," said Lozano, who imagined that the late Wilder Penfield may have felt the same way more than 50 years earlier when he used electrodes to find the target of seizures in his epilepsy patients. When Penfield sent electricity through the electrodes, it would provoke memories, emotions and even specific behaviors depending on where the electrode was placed. This enabled him to map the human motor cortex. "It is Penfield revisited," Lozano said. "It is like going where no man has gone before and figuring out what is taking place in the mind."

With Deanna, as with previous patients, Mayberg was struck by these instant responses. "When we started, we didn't know what to expect," she said. "We hoped there would be an antidepressant effect over time, but we had no expectation of any acute effects—certainly not the types of changes described in the operating room. Patients who had the effects described a sudden sense of relief and calm as they became aware that their unrelenting negative mood had suddenly changed. With many more patients we have learned that while these acute effects are extremely interesting, not all patients experience them, and [those who do not] still do well with long-term DBS. That said, this was one of those aha moments that we could not have imagined. It has given us many new ideas to test about the nature of antidepressant mechanisms."

Another patient who had been deeply depressed for years spoke poetically. "Did you just do something?" Mayberg recalled the patient asking. "I have this sudden sense of calm—the difference between a laugh and a smile. Like the first day of spring when you see the crocuses peeping through the snow."

The team later received approval to expand the study, first to a sixth patient and then to 20, and they witnessed similar reactions from many who received the experimental treatment. Mayberg began to wonder just what was going on. "Who in their wildest dreams thought that this would be the phenomenon?" Mayberg said. Besides the patients' feelings of relief, there was a renewed social engagement, a feeling of being awake and aware. "They are paying more attention to us," she said. But ultimately the researchers didn't know whether these initial and surprising responses meant anything for the bigger picture. Would patients get better?

Mayberg found the entire experience "scientifically inspiring" and says she now realizes that further work on deep brain stimulation is necessary. "It's hard not to feel exuberant about the notion that these seriously ill patients might now have access to a procedure that could actually transform their lives," Mayberg says, years into studying deep brain stimulation as a treatment for depression. "This is highly selective modulation. It gives us important clues where to shine the light."

It takes time for depression to work its way into the brain networks to cause symptoms, and the return from it is often slow going for patients—even with the stimulators, doctors now agree. The results from the first six Toronto patients showed a two-thirds response rate at six months; this effect has been maintained in these patients after more than four years of continuous DBS. So far there have been no long-term side effects with the target that Mayberg and Lozano are using in the DBS procedure and, at the doses used, batteries are lasting a minimum of four years. Some patients in the studies have had infections at the site of the electrodes or problems with the leads or the device that forced them to have the device replaced, but they responded to treatment once these short-term problems were resolved.

Mayberg moved to Emory University in 2004 and continues to study deep brain stimulation in severely depressed patients, funded in part by the Dana Foundation. She maintains a working relationship with the Toronto team.

On a parallel track, scientists at Brown University, Cleveland Clinic and Massachusetts General Hospital are also studying deep brain stimulation for depression. This tight-knit collaboration consists of Brown psychiatrist Benjamin Greenberg, Cleveland neurosurgeon Ali Rezai and Mass General psychiatrist Darin Dougherty, as well as a team of other psychiatrists, neurosurgeons, psychologists, nurses and technicians. They have identified a different target—the ventral anterior limb of the internal capsule—that Rezai said is on the same avenue as Mayberg's area 25 but on a different block. Their idea to use DBS in depression followed on earlier work that demonstrated its success in treating symptoms of obsessive-compulsive disorder (OCD), Greenberg and Rezai said. Many OCD patients also suffer from depression, and those symptoms seemed to get better after the device was turned on. The group at Brown University began performing deep brain stimulation on patients with obsessive-compulsive disorder in 2000, and it was only a matter of time before their team went after depression. "The first thing we saw with our OCD patients was a change in mood," said Rezai, who joined the Brown University team in 2001. "They reported that the weight of the world had lifted off their chests." Rezai saw similarities in depressed patients and wondered whether the technique could alter primary depression as well.

One of the first patients the team operated on was Diane Hire, a 54-year-old woman with a 10-year history of unrelenting depression. She was so despondent that she barely spoke. Suicide was constantly on her mind, and she had attempted to end her life several times. She had tried every medicine known to tackle depression and had had more than 70 sessions of electroshock therapy.

When Rezai turned the stimulator on in the operating room, Diane immediately brightened and laughed. She said recently that she hadn't smiled in a decade. A day later, as scheduled, she had the battery pack implanted in her chest wall.

A week later the stimulators were turned on. Since then she has settled into the normal rhythms of everyday life, which is replete with a mix of emotions. But now her emotions are appropriate to her experience. In the process, she has also dropped 120 pounds that she had packed on during her depression. In 2008, Hire told her story at a conference on DBS and said of the treatment: "I wake up every day happy to be alive. I wake up looking forward to what is ahead. I am who I was. I am not a new person or a changed person. I am who I was."

The future of DBS research

In April 2008 Medtronic, the Minneapolis-based company that manufactures a deep brain stimulator called Activa, announced plans for a DBS depression study, in which Greenberg, Rezai and their collaborators are participating. Elsewhere, Advanced Neuromodulation Systems (ANS), a division of St. Jude Medical Center in St. Paul, Minn., is enrolling patients in a multi-center clinical trial to test its stimulating device, called Libra. Mayberg and Lozano have a patent licensed by ANS and will consult on the study. ANS will go after area 25. Both studies will accept only the sickest depressed patients, those for whom nothing else has worked. Investigators will pursue the various targets they have previously identified.

Apart from these new studies, in the first leg of experimental trials, more than 60 patients in the United States, Canada and Europe have had stimulators implanted into their brains to treat depression; a few years after the initial attempts, DBS is still very much an experimental procedure. The modern-day pioneers of experimentally treating patients with electrodes implanted in the brain—the Brown/Cleveland/Mass General team and the Toronto and Emory groups—are watching this first generation of patients very closely. "We want to treat depression like we treat heart disease," said Mayberg. "If something goes wrong, it's over. We are being very careful."

Researchers still have much to learn about deep brain stimulation for depression: How can the settings be optimized? Is one-sided stimulation adequate? How does DBS actually work? Finding answers will require ongoing research and dialogue between academic researchers and industry.

Today a high level of checks and balances is in place, including those required by the U.S. Food and Drug Administration, which oversees new technologies and treatments, and those of individual hospital institutional-review boards. Scientists trying to find ways to alter the landscape of the human brain are treading slowly and very carefully.

CHAPTER 4

A Road Paved by Reason

By Elizabeth Norton Lasley

Elizabeth Norton Lasley is a writer with a specialization in neuroscience. Formerly a senior editor at Dana Press, her freelance articles have appeared in numerous publications, including Science.

In 2007, Dr. Aaron "Tim" Beck of the University of Pennsylvania won the prestigious Albert Lasker Award for Clinical Medical Research for the development of cognitive therapy. Cognitive therapy is one of the few forms of psychotherapy that has been rigorously tested in clinical trials. It was first developed to treat depression, but its benefits extend to obsessive-compulsive disorder, post-traumatic stress disorder and perhaps even such "physical" ailments as hypertension, chronic fatigue syndrome and chronic back pain.

PSYCHOLOGICAL PROBLEMS result from the erroneous meanings that people attach to events, not from the events themselves.

This central principle of cognitive therapy, identified by Aaron "Tim" Beck in the early 1960s, has provided no less than a basic structure for understanding human nature, particularly with respect to emotional disorders. Today Beck, 87, is among those considering new applications for cognitive therapy even as he receives recognition for its development and use in mood disorders such as depression.

According to the central principle, depressed patients, for example, interpret their experiences in terms of their sense of failure and helplessness. Patients with anxiety think in terms of threats to their physical well-being or social acceptance. People with obsessions perceive their thoughts as dangerous, disgusting or immoral.

In cognitive therapy, patients learn through a variety of strategies to test their faulty beliefs. They then learn to appraise themselves and their futures in a way that is realistic, unbiased and constructive.

Beginnings

Although cognitive therapy has proved highly successful at improving the lives of people with depression and many other conditions, this success would not have been possible without a detailed understanding of what goes on in these patients' minds. Indeed, Beck feels that his greatest contribution was to delineate the inner workings of depression. His work with his own patients, plus his persistent questioning of then-available methods, touched off a chain of reasoning by which he worked out the process of depression in the troubled mind. He found that people who are depressed systematically block out the positive aspects of their life, seeing only the negative. They interpret ambiguous events in a negative way, which he describes as cognitive distortion. If something genuinely negative does occur, they tend to exaggerate its magnitude, significance and consequences. A minor error becomes a major catastrophe. A normal problem becomes an insoluble dilemma. The result of such negative thinking is that the individual feels sad and hopeless, withdraws from other people and may become suicidal.

"I was privileged to start my research on depression at a time when the modern era of systematic clinical and biological research was just getting under way," says Beck. "Consequently, the climate was friendly for such research, and the field for new investigations was wide open." It was the late 1950s. The National Institute of Mental Health had only recently begun funding research and providing salary support for full-time clinical investigators. A national organization called the Group for Advancement of Psychiatry, under the leadership of professionals who were dissatisfied with the field's lack of scientific underpinnings, was providing guidelines, as well as the impetus, for clinical research.

Caught up in the spirit of the times, Beck was prompted to start his own line of research. He was particularly intrigued by the paradox of depression. The disorder appeared to violate the time-honored canons of human nature: the self-preservation instinct, the maternal instinct, the sexual instinct and the pleasure principle. All of these normal human yearnings were dulled or reversed by depression. Even vital biological functions like eating or sleeping were attenuated.

Beck had a "eureka moment" in the early 1960s while reviewing findings from his clinical research in depression. "In talking to my wife about this, everything seemed to click into place all at once." The various components of Beck's research came together: the discovery that negative beliefs shaped his patients' interpretations and that these negative interpretations (or cognitions) then led to the sad feelings, social withdrawal and, especially, suicidal wishes. When the beliefs and cognitions were modified in therapy, the distorted interpretations and the symptoms of depression diminished. Because the distorted cognitions became the focus of treatment and the process of change depended on cognitive restructuring, Beck applied the label "cognitive therapy" to the treatment.

Tools for Patient and Therapist

Although cognitive therapy usually focuses on problem solving in the present, by doing that task the patients also develop lifelong skills. With the therapist's help, they learn to identify distorted thinking, modify beliefs, relate to others in different ways and change their behavior. Early in the process the patient sets goals for improving relationships, work, moods and symptoms. Other desired areas of improvement might include pursuing spiritual, intellectual or cultural interests; increasing exercise; decreasing bad habits; or learning new interpersonal or management skills, at work or at home.

In addition to providing strategies to re-pattern negative cognitions, throughout the 1960s Beck developed a number of scales to specifically measure depression, anxiety and suicidality. These are based on the patients' descriptions of their symptoms, feelings, thoughts, wishes and behaviors. The descriptions are converted into items or questions, and each is given a numerical weight. The total scores are then correlated with the clinician's evaluations of the severity of the particular illness. These scales have proved to be effective tools in measuring the extent of the patient's disorder.

Beck's scale that is used for evaluating suicidal behavior, for example, not only predicts who is most likely to make another attempt at suicide but can offer a means of prevention. Called the Beck Hopelessness Scale, it details the individual's evaluation of his or her future. In a 20-year prospective study published in 2000, and a 30-year study now in press, Beck found that a high initial score on the hopelessness questionnaire is a reliable predictor of who will commit suicide. The information has been used successfully to intervene in suicide: working with patients who have previously attempted it, Beck and colleagues have found that cognitive therapy specifically targeting hopelessness can reduce the likelihood of a subsequent attempt by almost half, as well as reducing both hopelessness and depression.[1]

As Good as Medication, or Better

In terms of treating depression, Beck says some studies show that cognitive therapy can be just as effective as pharmacotherapy and superior in preventing relapse. For example, a study in the April 2005 Archives of General Psychiatry found that the benefits of cognitive therapy endure well beyond the end of treatment.[1] Patients first underwent 16 weeks of treatment with either cognitive therapy or antidepressants. Those who responded well to each regimen entered a second, yearlong phase of the study. At the beginning of this phase, patients who had received cognitive therapy stopped their treatment, except for just three "booster" sessions throughout the year. Patients who had received medication were randomly assigned either to continue their medication or to be switched to a placebo on a double-blind basis.

The results reported in the study: only 30 percent of patients who concluded cognitive therapy relapsed into depressive symptoms, compared with 76 percent of the patients withdrawn from medication and 47 percent of those continuing with medication.

The authors speculated that the lasting effects of cognitive therapy reflect the patients' newfound ability to "do the therapy for themselves." They remarked that the strategies learned "eventually become second nature, coinciding with a parallel change from problematic underlying beliefs to more adaptive ones." In this way, the patient is less likely to become distressed in situations that previously would have spiraled into a depression-producing pattern of thought.

In 2006, Beck published a meta-analysis of more than 100 studies that found similar success for cognitive therapy when compared with medication.[2] In a study published online March 5, 2008, in the Journal of Clinical Psychiatry, a team of researchers screened more than a thousand studies of cognitive therapy used to treat anxiety disorders. Narrowing the field to

27 randomized, placebo-controlled trials, the authors found that cognitive therapy yielded significantly greater benefits than placebo treatment and may be even more successful when combined with pharmacotherapy.[3]

Brain Basis for Cognitive Therapy

Several imaging studies of cognitive therapy used to treat phobias, obsessive-compulsive disorder and anxiety show restored balance of brain activity in areas often over- or under-activated in patients with these conditions. A neuroimaging study of depression published in the January 2004 Archives of General Psychiatry found that cognitive therapy and pharmacotherapy bring about similar changes, but through different pathways.[4]

Antidepressants are often described as working from the bottom up. They adjust the exchange of chemical messengers at the synapse, the point of connection between neurons. With balance restored among the various chemicals—typically serotonin, dopamine and norepinephrine—a chain of events begins that ultimately results in the depressed patient's beginning to feel better. Exactly what goes on in the brain is not well understood, and the process takes some time. Most patients are on medication for at least three weeks before noticing a difference in mood.

Cognitive therapy, on the other hand, works "top down." Patients learn to monitor, question and redirect their negative interpretations of events—thus bringing much of their emotional state under conscious control. The resulting improvement in mood has ramifications throughout the brain, presumably restoring balance in many specific aspects of the brain's functioning.

Helen Mayberg and colleagues, then at the University of Toronto, used positron-emission tomography (PET) to study the brain changes brought about by cognitive therapy. In a study published in 2004, 14 patients were scanned before and after they completed 15 to 20 sessions of therapy. The scans were then compared with published imaging studies of the brain in depression, both with and without antidepressant treatment.[4]

The results suggest that "top down" and "bottom up" are not mere metaphors. Cognitive therapy produced changes in several parts of the prefrontal cortex, the brain's topmost layer. These areas handle "higher" functions such as working memory, processing of personally relevant information and "cognitive rumination." These functions tend to be impaired in people with depression, and imaging studies often show abnormally high activity in these regions. Cognitive therapy decreased the activity in the prefrontal areas, suggesting improved functioning. On the other hand, the therapy caused increased activity in other areas deeper or "lower" in the brain: the anterior cingulate, involved in directed attention and monitoring of emotions, and the hippocampus, a nexus of memory encoding and consolidation.

The authors of the Toronto study postulated that the pattern of brain changes represents the neural counterpart of cognitive therapy. As patients learn to observe their emotional responses to life events, block the automatic resurgence of distressing memories and reduce their tendency to brood and overanalyze irrelevant information, the relevant parts of the brain return to a balanced level of activity.

Antidepressant treatment affected many of the same areas but in mirror-image ways—decreased activity in the memory- and attention-serving areas such as the hippocampus and cingulate, and increased activity in the frontal regions that help bring thoughts, and possibly feelings, under conscious control. The finding supports the idea that the treatment approaches work in complementary ways—cognitive therapy from the top down, and medication from the bottom up—ultimately stabilizing a complex pathway running between the hippocampal and prefrontal areas.

These visible, measurable effects on the brain may provide one of the best answers to the criticisms that are aimed at cognitive therapy. Among those who sound a cautionary note are members of an original "bottom-up" school of thought: psychoanalysis. In the Freudian tradition, emotional disorders result from past traumas carved so deeply into the psyche that the patient is unaware of them; curing the disorder means excavating these conflicts from the unconscious through the lengthy process of psychoanalysis. Cognitive therapy de-emphasizes the importance of circumstances and events, putting the conscious mind firmly in control.

Peter Fonagy, a professor of psychoanalysis and head of the Anna Freud Center in London (a center for treating children and families with psychological problems), does not suggest that cognitive therapy is ineffective. He does, however, argue that psychoanalysis remains a valuable tool for many patients. Citing studies showing that the benefits of cognitive therapy in anxiety disorders fade over time, Fonagy told the British magazine Prospect that cognitive therapy is "marketed as an antibiotic when it's really an aspirin."

In the May 2003 American Journal of Psychiatry,[5] psychiatrist Gordon Parker of the University of New South Wales, Australia, and colleagues reviewed many of the studies showing the superiority of cognitive therapy over other forms of psychotherapy. Again, although the authors did not question the effectiveness of cognitive therapy, they did question the designs of studies that showed it to be universally applicable. Too often, the authors observed, psychotherapies (and medications, too) are tested in a group of patients whose disorder is classified according to its severity—not according to how it manifests in patients, which may be affected by causes ranging from ongoing life difficulties to poor interpersonal skills to biochemical imbalance.

Arguing that cancer treatment modalities, for example, are prescribed according to the nature, not the severity, of the patient's cancer, Parker and his co-authors contend that different types of treatment, including medication, may be warranted for different types of depression. To accurately assess whether cognitive therapy is superior—and for whom—the authors suggest studies of the patients that do respond well, not just those who improve initially but those who continue without relapsing. Pinpointing the "responders" would draw a clearer picture of the success and limitations of cognitive therapy—providing a targeted approach to treating psychological disorders.

Practitioners of cognitive therapy are open to incorporating tools from other psychotherapeutic approaches, such as psychoanalysis, and treatment with cognitive therapy can and often does include medication. In Beck's view, brain imaging studies offer the best counter to claims that cognitive therapy doesn't get at the "real" cause of psychological disorder. "The changes that cognitive therapy produces in key brain regions, and the more enduring effects compared to pharmacotherapy, would not happen if the therapy simply addressed the symptoms. It must be getting at the causes," says Beck. The brain is known to rewire itself according to experience, in a process known as plasticity. Thus the changes brought about by treatment—observable in imaging studies—can literally reconfigure the old circuitry that would otherwise have continued the depression-producing thought processes indefinitely.

Future Directions for Cognitive Therapy

Beck is not done with considering how cognitive therapy works and how else it could be applied. Its principles are offering hope not just for mood disorders and mental illness but also for a number of conditions not typically considered "psychological." Beck is conducting a trial of cognitive therapy to ameliorate the "negative" symptoms of schizophrenia, such as apathy and social withdrawal (as opposed to what scientists term "positive" symptoms, such as auditory hallucinations, which are still treated with medication). The therapy has also proved useful in treating medical conditions resulting from the two Gulf Wars, such as post-concussion syndrome and post-traumatic stress disorder.

Beck believes that, in addition to effectively treating the classic psychiatric disorders, cognitive therapy will be increasingly used to treat more "medical" problems such as hypertension, chronic fatigue syndrome and chronic back pain. "A variety of dysfunctional cognitive reactions can lead to medical conditions," he notes. "For example, an inappropriate level of anger can aggravate hypertension. Excessive attention to physical feelings, and exaggerated interpretation of their significance, can play a role in chronic fatigue syndrome and lower back pain. In general, negative thoughts and distorted interpretations can exacerbate almost any physical symptoms." Cognitive therapy is increasingly used, often in conjunction with medication, to both relieve the stress that aggravates these disorders and enhance the patient's adherence to the more "medical" end of the regime.

Beck is also researching the neurobiological correlates of the cognitive model in depression, with an aim toward pinpointing the changes that cognitive therapy produces in the brain in a variety of disorders. He expects to see further integration of cognitive therapy with other psychotherapeutic approaches that have been proved to be valid: "I doubt that there will be as much fragmentation in the psychotherapeutic field. In all likelihood, there will be one psychotherapy incorporating a variety of approaches, depending upon the patient's characteristics and the nature of the disorder."

Beck, who directs the Center for the Treatment and Prevention of Suicide, based at the University of Pennsylvania, and works with city-run mental health organizations throughout Philadelphia, also anticipates greater dissemination of cognitive therapy into the community.

He views with satisfaction the widespread acceptance of cognitive therapy in institutional and community settings: "Various managed-care companies and mental health centers now expect their therapists to be trained in cognitive therapy. The British government has recently set up a large program for training over 6,000 mental health workers to do cognitive therapy. There are now dozens, if not hundreds, of researchers focusing on the theoretical underpinnings of cognitive therapy, or on its applications."

In the end, Beck returns to the continuing promise of cognitive therapy for fields in which it has already proved so effective. "One of the most promising directions for the present and future is in preventing mental disorders from taking hold," he concludes. "Several studies have been done, and others are under way, to identify those at risk for depression and suicide. With the tools of cognitive therapy, early intervention can help prevent negative thought patterns from developing into full-blown mental illnesses."

Interpersonal Therapy

Another form of psychotherapy was inspired by Aaron "Tim" Beck's early efforts to hold cognitive therapy up to the same rigorous scrutiny as pharmacological interventions. In the 1970s, psychiatric epidemiologist Myrna Weissman and psychiatrist Gerald Klerman, both then at Yale University, developed what came to be known as "interpersonal therapy" as part of the first large-scale clinical trial using both drugs and psychotherapy to treat depression.

Interpersonal therapy is based on the premise that depression often occurs along with the onset of a major life event involving relationships—such as ongoing difficulties with a spouse, friend, co-worker or family member; the loss of a loved one; or the inability to form close attachments. In interpersonal therapy, the patient and therapist agree at the outset on an appropriate length of time for their work together—anything from a few weeks to more than a year, with a few months being typical. They then address depression specifically as it manifests in the patient's life situation and relationships. Rather than focusing on thought processes about an event, the therapist will explore what led to the problem—disputes in job or family relationships, for example—and work out strategies to either improve the situation or move on from it. With better coping tools in place, the patient is more likely to respond to future problems in productive ways that do not lead to depression.

Klerman strongly emphasized the need for psychotherapy to mimic clinical practice. Beck used his own not-yet-published manual, Cognitive Therapy of Depression, widely in his clinical trials. But most patients with depression received various untested therapies lumped together as "supportive therapy." Klerman advocated describing interpersonal therapy with the same clinical rigor that Beck had used in his manual.

"We felt strongly that we had to test the efficacy of interpersonal therapy before advocating its widespread use," says Weissman, who is now at Columbia University (Klerman died in 1992). Throughout the 1980s, she and others conducted their own clinical trials. Their manual was published in 1984 and has been revised several times; 2007 saw the publication of the Clinician's Quick Guide to Interpersonal Psychotherapy, from Oxford University Press.

"Practitioners of interpersonal therapy share a close affiliation with Dr. Beck," Weissman notes. "Both therapies are time-limited, although the limits can vary from a few months to several years. Both have guidelines strictly set down in manuals, have a strong diagnostic component and are designed to overcome depression."

Cognitive therapy and interpersonal therapy remain the most widely used therapies for depression, Weissman says—and not only in affluent, industrialized countries. "We got a call one day from Paul Bolton at Johns Hopkins University saying that World Vision International [a nongovernmental humanitarian organization] was interested in treating depression in Uganda. Dr. Bolton wanted to use interpersonal therapy," says Weissman. "It was the most interesting request I'd had in a long time."

Trials of interpersonal therapy in Uganda, a country ravaged by war, poverty and AIDS, have shown the approach to be remarkably effective, especially among women. Specifically, Uganda has had notable success in reducing the incidence of AIDS. Health workers there feel that combating depression is an important aspect of this success, since depressed people often engage in risky behavior. With the intensive involvement of local health workers, Bolton and Weissman, along with Helen Verdeli and Kathleen Clougherty of Columbia, tailored interpersonal therapy to a uniquely African setting—taking into account many differences in communication styles. For example: "When a woman in Uganda is angry at her husband, she doesn't chew him out. She cooks him bad food," says Weissman.

In an initial small clinical trial in Uganda, the results of which were published in 2003 in the Journal of the American Medical Association, interpersonal therapy proved highly effective in reducing depression: after therapy only 6 percent of the treated group met the criteria for major depression, compared with more than half of the untreated control group.[6]

A 2007 study found similar results with another high-risk Ugandan group—teenagers displaced by war to refugee camps.[7] Again, the difference was more striking among women. The reason is unclear, but Weissman suspects that alcoholism, much more prevalent in men, may play a role. She adds that interpersonal therapy and cognitive therapy are used all over the world, with texts translated into numerous languages. An international society of interpersonal therapy will be meeting in New York in March 2009.

CHAPTER 5

The Political Brain

By Geoffrey K. Aguirre, M.D., Ph.D.

Geoffrey K. Aguirre, M.D., Ph.D., is an assistant professor of Neurology at the University of Pennsylvania. He studies the neuroscience of higher-level visual function and uses brain scanning techniques to determine neural information processing when subjects view faces and other objects.

Research using neuroimaging to detect the emotional response of undecided voters has led to controversy among scientists. An op-ed article in the New York Times, written by the leader of one such study, argued that brain scans could help determine the voters' true feelings about candidates, eventually making pollsters obsolete. Dr. Geoffrey Aguirre discusses the flaws of this argument, the feasibility of this method to determine hidden preferences and the ethical issues inherent in the process.

BY NOVEMBER 11, 2007, the Democratic and Republican presidential nominating contests were well under way. The Democratic candidates spoke that night at the Jefferson-Jackson fundraising dinner in Iowa, and a second debate was approaching for the Republicans. With the first votes of the caucuses and primaries only weeks away, pollsters and pundits were working to divine the intentions of voters, particularly the coveted "swing" voters not committed to a candidate. Which Republican would appeal to women, closing the so-called "gender gap"? Was anyone truly undecided regarding Mrs. Clinton, a candidate who had been in the political spotlight for more than 15 years? That Sunday, the op-ed page of the New York Times promised insight into these central questions, in the surprising form of pictures of brain activity.

Neuroscientists from the University of California, Los Angeles, led by Marco Iacoboni, had used functional magnetic resonance imaging to measure the responses of undecided voters to the candidates. Their conclusions were startling in their depth and breadth. One Republican candidate, Fred Thompson, was found to evoke particularly strong feelings of empathy. Further, while some voters said that they disapproved of Hillary Clinton, their brain activity revealed that they had unacknowledged impulses to like her. The study had seemingly reached into the minds of voters and plucked out their hidden emotions and conflicts. Perhaps political talk-show hosts and Gallup pollsters would soon be unnecessary. Why analyze and poll when the feelings and intentions of voters could be read directly from their brains?

Instead of sparking a revolution in political science, however, the editorial provoked broad condemnation from the neuroscience community. Within days the New York Times had published a letter from 17 scientists who argued that the study was fundamentally flawed. At scientific meetings and on the discussion boards of Web sites, the hue and cry continued. The prominent scientific journal Nature published a scathing editorial[1] that lamented the absurdity of the study. After more than a decade of increasing publicity for brain-scanning results in the lay press, the Iacoboni editorial had provoked a backlash. Neuroimaging had jumped the shark.

For his part, Iacoboni defended his study. In an online letter, he argued that the approach he used in his study of voters is common to many cognitive neuroscience experiments. If all those previous studies were valid, he asked, was his study considered flawed simply because he had left the ivory tower to examine political candidates or because he reported his results in a newspaper? Iacoboni's defense raises challenging questions for scientists and consumers of scientific studies. If his group's undecided-voter editorial column is flawed, are there scientific studies that use comparable methods, published in respected, peer-reviewed journals, that are also absurd? What, exactly, was so wrong with his study, given that it used modern neuroimaging techniques and analyses? Could there be valid studies of political topics that would either provide insight into political thought or be of value to a pollster or candidate? To address these questions, we must first understand how raw neuroimaging data can be transformed into a picture of brain activity that a researcher might interpret as showing latent sympathy for Hillary Clinton.

Brain Imaging Approaches

Magnetic resonance imaging (MRI) has been used for some decades to construct pictures of brain anatomy. Functional MRI (fMRI), developed in the 1990s, offers a measure of brain activity. For fMRI data to be collected, a participant lies on a table that is slid within a powerful magnet. The subject receives instructions and is presented with pictures and sounds during the scan. Meanwhile, weak radio waves are used to measure the effect that nerve cell activity has on the magnetic field. The effect is indirect; local changes in brain activity induce a cascade of effects on blood flow, on oxygen, and in turn on the iron atoms in hemoglobin molecules that ultimately warp the microscopic magnetic field. The procedure is extremely safe and painless, and it can be completed in about an hour. Nerve cell activity can be measured over the entire brain from second to second, with millimeter resolution.

An image of brain activity is not available immediately after the scan. To create a picture, a researcher must first decide which two (or more) behavioral conditions are to be compared. This is an important, and generally unrecognized, aspect of neuroimaging studies. There is no brain picture "for" anxiety or memory. Instead, the experiment must compare the relative brain activity between two behavioral states, with the hope of isolating the mental operation of interest. To study anxiety, one might present the subject with pictures of snakes and guns and then at another time show pictures of puppies and flowers. The experimenter might conclude that a brain region, such as the amygdala, that shows a greater neural response to the snakes than the puppies is responding to the differential anxiety provoked by the stimuli. The colorful brain image simply shows where statistically greater activity was seen for one condition as compared with the other.

This approach to brain imaging, in which the experimenter tries to manipulate the mental state of a subject in order to then observe the evoked brain activity, is termed "forward inference." Experiments like this dominated the application of neuroimaging for many years. The study of sensory processing has been particularly successful, in part because the mental states to be studied can be differentially evoked quite readily. For example, a brain region, "area MT," has been identified that invariably responds when the subject sees something moving but does not respond to static pictures. Neuroimaging and forward inference have been used to study more complex behavioral states as well, such as emotion, conflict resolution, sense of self and reward processing. Specific brain areas have been found that reliably increase their neural activity during these behaviors, although the link between a particular behavior and a brain region is more tenuous. First, it is challenging, and in some cases arguably impossible, to compare two complex behavioral states and leave behind the isolated mental concept of, for example, greed or risk-taking. These behaviors are necessarily embedded in complex tasks and emotions and cannot be isolated by experimental design in the same way that visual motion can be. Second, the attempt to map a single behavior to a single brain region quickly breaks down past early sensory representation. The amygdala may consistently respond more strongly to anxiety-provoking stimuli, but it is also activated by positive stimuli (puppies and flowers) as compared with neutral pictures (toasters and trees). The state of affairs is even worse for areas of the frontal lobe, where dozens of different mental operations have been identified that might activate a given square centimeter of cortex. A related complication is that different subjects may have quite different behavioral or emotional responses to a particular experimental situation, foiling attempts to describe a consistent relationship between behavior and brain region for a population.

The application of neuroimaging to political questions does not involve "forward inference," however. Political neuroimaging, along with the burgeoning fields of social, economic, and even marketing neuroscience, relies upon the opposite approach. Instead of determining the brain region associated with a particular behavioral state, a "reverse inference" study attempts to identify the behavioral state of subjects by observing their brain activity. Initially, studies of this kind examined basic sensory phenomena. The activity within the aforementioned area MT might be used to determine if a particular optical illusion induces a sense of motion in some people. Such a conclusion could be well supported. After dozens of "forward inference" studies, it has become quite clear that the perception of motion, and only motion, is always associated with activity in this patch of cortex. The reverse inference approach has also been used to probe more complex behaviors. Activity within the insula when a subject is presented with recognizable lies has been taken as evidence that lies induce the same sense of disgust that rotten food does, as the latter has also been observed to activate the insula.

The Trouble with Reverse Inferences

The problem, of course, and the source of the widespread displeasure with Iacoboni's newspaper article, is that these reverse inferences are only as good as the evidence that supports a unique mapping of a particular mental operation to a particular cortical region. And for many of the claims that Iacoboni makes, this evidence is not good at all. The presence of an amygdala response to pictures of Mitt Romney did not necessarily indicate anxiety regarding his becoming president, as positive emotions can activate this region as well. A further limitation is that the response to pictures of Mr. Romney was compared with (presumably) the neural response elicited by a blank screen. The amygdala response may have been not to Mr. Romney per se but to his attractive hair. Finally, even if we were to grant that amygdala responses indicated anxiety, and were specific to Mr. Romney himself, perhaps the subject was simply anxious because his favorite candidate, Mitt, was not doing well in the polls!

Further compounding these weaknesses is Iacoboni's tendency to engage in what might be termed "neuromythology." When presented with a picture of a brain with colorful activity, he has a tendency to spin a yarn to explain what he sees. The claim that voters who stated a dislike for Mrs. Clinton actually harbored latent kind feelings toward her was not even partially implied by the faulty logic of the study; rather, it was an explanation, made up from whole cloth, for the observation of cortical activity that implied "conflict." This unfortunate tendency to treat neuroimaging data as a Rorschach blot is on full display in a recent article in the Atlantic in which the author, Jeffrey Goldberg, visits with Dr. Iacoboni and his associates who operate a "neuromarketing" company. The initially uncomfortable finding that Mr. Goldberg had a "positive, reward" response to a picture of Mahmoud Ahmadinejad leads to the tortured explanation that the author is actually imagining the happy day that the Iranian president is deposed. Equally bereft of logic is the explanation of how the equivalent responses of Mr. Goldberg's brain to Hillary Clinton and his own wife actually signify two quite different behavioral states.

Does the preceding criticism suggest that a valid study of political behavior using neuroimaging is not possible? No. Instead, while there are pitfalls to be avoided, much might be learned regarding the behaviors and emotional states that people develop and deploy in evaluating political candidates. To be successful, such studies must compare carefully controlled states to isolate a behavior of interest and draw well-supported inferences regarding the activity seen. In fairness, Iacoboni and his colleagues have published an example of such a study.[2] Beyond simply being valid, however, there is an additional requirement that a neuroimaging study of political behavior be useful: it must provide an insight not available by simply asking a voter his or her opinion.

Imaging Versus Polling

For the most part, human behavior is readily available to be observed or queried. It would not come as a surprise to learn that voters who identify strongly with one party tend not to like candidates from the other party. Thus, it seems an unnecessarily roundabout way to learn this truth by measuring increased amygdala and insula responses to pictures of opposing candidates. Similarly, if you want to know how someone will vote for a candidate, you can generally just ask the person. The chief challenge for pollsters is obtaining a sample of responses that are representative of the population, a problem that would not be solved by neuroimaging. There is nothing automatically more informative about measuring neural activity as compared to directly observing behavior.

There are many circumstances, however, in which asking voters their opinions will not provide the entire story. In the face of an overt desire to mislead or a simple lack of introspection, neuroimaging of political behavior might provide insights not otherwise available. For example, a plausible study might examine the emotional response to political "spin." Politicians frequently provide an unrealistically favorable description of events, omitting details that are inconvenient. While voters claim that they object to spin, they may nonetheless respond positively. Given previous studies that have identified patterns of brain responses for overt lies as compared to truths, what is the response to spin? Is spin treated as a lie, and how is this modulated by one's political affiliation? There are certainly many other topics in the realm of political behavior that fall into this category and could eventually come under study.

We may also consider applications of neuroimaging techniques to assist polling in cases where voters are unwilling or unable to provide accurate responses. Obviously, a source of much uncertainty in polling results is "undecided voters." Perhaps some proportion of voters really do have a strong preference but are insufficiently confident to share this with a pollster. Further, voters may consider one candidate to be the more socially acceptable choice to report to the pollster, although they intend to choose the other in the privacy of the voting booth. This is the "Bradley effect," named for Tom Bradley, an African American former mayor of Los Angeles who lost his 1982 race for governor despite polling that showed him ahead of his white opponent.

Could neuroimaging be used to determine true voting preference in these cases? Perhaps, although not in any straightforward way. Simply presenting the candidates' pictures and recording a response would not be enough. As we have considered, the presence of, for example, an amygdala response to one candidate cannot be taken as evidence that the voter will vote a certain way. Recently, techniques to analyze the pattern of neural responses across the entire brain have been developed. These "multi-voxel patterns" (MVPs) can be used to deduce a subject's unstated intention in controlled settings. For example, if a subject is presented with two targets on a screen and told to choose one but not yet indicate which, the choice can be accurately read from the MVPs in advance of the response. It is possible that the pattern signature for responses for a given voter could be measured while the person is making a series of innocuous decisions. In the critical test, the subject would then be presented with pictures of the candidates, side by side. Although the voter would withhold an overt response, the implicit preference might be available in the distributed fMRI data.

Suppose that this were shown to be a valid way to measure implicit voter preference—would it be of practical value? Only a small number of subjects could ever be examined in this fashion, as the collection of such data is a time-consuming and expensive undertaking. Further, obtaining a representative sample would be very difficult, as older subjects, for example, generally find it hard to participate in an hour-long, uncomfortable neuroimaging scan. Finally, simple polling questions and adjustments are available to address these challenges. Undecided voters can be asked to indicate which way they "lean," which predicts well how they will ultimately vote. The magnitude of the Bradley effect can be estimated by asking a voter if she thinks her friends and acquaintances would be hesitant to vote for a certain candidate, even if she professes to have no such qualms. Indeed, a recent paper in the journal Science[3] has demonstrated that purely behavioral techniques can be used to accurately predict the way an undecided subject will eventually vote.

Therefore, it seems unlikely that neuroimaging techniques will have much impact upon the practice of politics. Ultimately, politicians and political operatives care about behavior—if and how a voter will vote—and not much about the underlying neural basis for these actions. Simple polling provides this information much more readily and inexpensively than neuroimaging could ever do. In contrast, neuroimaging may find a place in the study of political science, in which the underlying motivations and behavioral states of voters have become an area of increasing interest.

Neuroimaging Our Preferences Versus Our Preference for Neuroimages

We have considered that neuroimaging techniques may be able, in principle, to identify voter preference. While this ability may be desired by politicians, it may be rejected by the polity. The secrecy of an individual's ballot is a cornerstone of modern democracy; if our voting preferences were known, we could be subject to the threat of retribution by a government we voted against. Fortunately, such an abuse of neuroimaging is unlikely. Given the size and noise of an fMRI scanner, no one could be scanned unknowingly. Moreover, an fMRI study requires tremendous subject cooperation, making these studies trivially easy to defeat.

While of little immediate risk, the possibility that neuroimaging might invade our political privacy has been of concern to ethicists who anticipate the impact of emerging neuroscience technologies. This attention is not inappropriate. It is almost certainly better for philosophers and ethicists to have their say before a technological revolution sweeps an unprepared society. I believe, however, that the attention and concern devoted to the possibility of a neuroimaging invasion of political privacy is somewhat misplaced. Greater and more immediate threats to privacy loom. In the same way that behavior in a laboratory setting or in a formal poll can accurately predict a voter's preference, so can our routine, daily actions provide a window to our intentions. Knowledge of where we live, what we buy, how we travel, and whom we know can be aggregated to provide information about our preferences. The possibility of this silent, creeping invasion of our privacy, advanced by profit-seeking corporations and terrorist-seeking government agencies, strikes me as far more menacing than the clanging of a seven-ton MRI scanner.

Instead of a threat to privacy, the principal risk is that misuse of neuroimaging will add further distraction and irrelevance to the political process. Although carefully designed neuroimaging studies might eventually provide valuable insights into political decision making, the slow, unglamorous grind of the scientific process will leave us time to be tempted by colorful pictures of the brain and stories of secret voter intention. The New York Times op-ed page is arguably the most influential two square feet of newsprint in American politics. The editorial column by Iacoboni and his colleagues stands as a testament not to the power of neuroimaging to make manifest our political preferences but to the manifest preference we all have for neuroimages.

CHAPTER 6

A Wound Obscure, Yet Serious

Consequences of Unidentified Traumatic Brain Injury Are Often Severe

By Wayne Gordon, Ph.D.*

Wayne A. Gordon, Ph.D., is Jack Nash Professor of Rehabilitation Medicine and associate director of the Department of Rehabilitation Medicine at Mount Sinai School of Medicine in New York City. Dr. Gordon has received a number of awards for his outstanding work in the field of traumatic brain injury.

* The preparation of this manuscript was supported in part by Grants H133A07033 and H133B040033 from the National Institute on Disability and Rehabilitation Research, United States Department of Education, Grant 1R49CE00171-01 from the Centers for Disease Control and Prevention and the generosity of the John Blair Haldeman Fund. The author wishes to acknowledge the constructive criticism of Margaret Brown, Ph.D., in the development of this manuscript.

Soldiers returning from war with visible head injuries are easy to spot, but what about soldiers—and civilians of all ages—who have brain injuries but no external wound? Wayne Gordon, Ph.D., notes that these cases of unidentified traumatic brain injury are far more prevalent than we realize. He offers suggestions for better awareness and treatment.

JOHN, AT AGE 3, WAS HIT ON THE HEAD by a swing at the playground. His mother called her pediatrician, who told her that she need not go to the ER because John had not lost consciousness. Immediately, her happy-go-lucky son seemingly became a different child: anxious and clingy. For a few years thereafter, John would occasionally shake his fists up and down, out of the blue, then stop; such episodes were later recognized as undiagnosed seizures. Initially, he did well academically but not socially. He became the butt of jokes and was labeled by his teachers as unmotivated and inappropriate.

Over many years, John's mother sought help from the schools he attended, his pediatrician, several neurologists, tutors and the like. No one was able to help. Finally, when John was an adolescent, a tutor told his mother that his reading problems were not typical and that he should be seen by a neurologist. The mother tried again. The boy was sent for a type of brain scan called single-photon emission computerized tomography, or SPECT, which showed major damage where the swing had hit his head 16 years earlier. However, the neurologist told her that there was nothing to be done; he was mistakenly of the opinion that it had been too long after John's injury for any intervention to be of use. John's mother persevered and found a program for him that could help address his cognitive and behavioral difficulties. Unfortunately, he was so emotionally damaged by so many years of being misunderstood—not only by everyone around him but also by himself—that despair won out. His traumatic brain injury ultimately ended in his suicide.

The brain injuries we see on the evening news, when soldiers return from war with visible, grievous wounds, are clearly evident: this is known traumatic brain injury, or TBI. Then (both in military and in civilian life) there are cases such as John's, where injury to the brain is relatively mild, with only a brief loss of consciousness or a period of feeling dazed and confused. Because the person appears physically unharmed, the "bump to the head" may easily be forgotten. This is appropriate in most cases, because the majority of people who experience mild brain injuries recover with no lingering effects. However, it is not appropriate for the large numbers of people who experience substantial post-injury cognitive, behavioral and/or emotional problems that do not go away. Unidentified TBI occurs when these problems are not understood to be a consequence of the head injury; they may be misattributed to aging or to stress or may never be explained at all. This type of unidentified TBI is a common phenomenon, one that needs attention from medical, educational and military systems—the last because TBI is "the signature injury"[1] of the wars in Iraq and Afghanistan.

The prevalence of unidentified TBI is difficult to determine both in civilian and in military populations because something that is not identified is, by definition, not counted. The best civilian estimates are based on extrapolations from the number of known injuries, which the Centers for Disease Control and Prevention place at 5.3 million (2 percent of the U.S. population).[2] The Centers acknowledge that these numbers underestimate the true prevalence of TBI, since only individuals treated in hospitals, those seen in ERs and those who die are counted. Not included are those who receive care outside of hospitals (e.g., in medical offices) or who do not receive medical attention at all (e.g., people injured in assaults, domestic violence, falls and the like). Research suggests that for every person hospitalized with a brain injury, three to five others are injured but do not receive any care.[3, 4] So the question is: among people who have sustained a brain injury, do we have any idea how many continue to experience symptoms commonly found after mild TBI but fail to causally link the symptoms to the injury? We have data that begin to answer this question. For example, in a population-based survey in New Haven, Conn., Jonathan Silver at New York University and colleagues at Columbia University found that 8.5 percent of the 5,034 people surveyed reported a brain injury with continuing challenges.[5] An unpublished study at the

Mount Sinai School of Medicine in New York City found a similar level of unidentified TBI: about 7 percent of a sample of people identifying themselves as non-disabled met criteria for TBI and also reported numerous symptoms associated with known TBI. If we consider such data in the context of the current U.S. population, they suggest that unidentified TBI may affect as many as 20 million to 25 million Americans. Clearly, more studies are needed to refine these estimates and get a better handle on the extent of the problem among civilians.

The number of unidentified TBIs in the military is also difficult to determine. We have learned in the past few years, thanks to media coverage, to expect large numbers of soldiers to have a known TBI. In reality, the prevalence of "probable" TBI is estimated at 19.5 percent, which translates to possibly 320,000 of those returning from Iraq and Afghanistan.[6] However, these numbers are probably underestimates because post-deployment screening has yet to be fully implemented, and many soldiers do not acknowledge their mental health challenges.

Widespread Consequences

The large number of estimated injuries in both civilian and military venues should raise concern, as the consequences of TBI, whether known or unidentified, can be life changing. TBI is strongly associated with multiple, often overwhelming challenges that can undermine a person's efforts to live a productive life, leading to "social failure." For example, among prisoners, estimates of the prevalence of TBI range from 42 percent to 87 percent;[7-9] for most, the brain injury preceded the start of criminal activity. TBI is also common in inpatient psychiatric and substance abuse populations, and, similarly, the injury often precedes onset of psychiatric symptoms[10,11] or substance abuse. Most of these brain injuries had gone unidentified prior to the respective studies. TBI is associated with high levels of co-occurring depression, anxiety and post-traumatic stress disorder, or PTSD.[6,12] (While some symptoms of TBI and PTSD are similar, such as fatigue and difficulty sleeping, other symptoms are unique to each disorder—heightened startle response and night sweats are unique to PTSD, for example.) Although the overlap between the estimated 320,000 returning soldiers with known TBI and the 300,000 returning with depression and/or PTSD is relatively small at 7 percent, this figure includes only those with identified TBI. In Silver's study, New Haven residents reporting TBI attempted suicide four times more often than those with no brain injury, they were more likely to be receiving public assistance or disability benefits and they experienced poorer overall emotional and physical health.[5]

Researchers are trying to determine what causes the extensive and often severe emotional and behavioral consequences of TBI, but they face many challenges in that quest. Linkages between injury site(s) and specific post-trauma symptoms have not been well established, and the lesions that occur following TBI are likely to be diffuse rather than localized. For so-called mild (and often unidentified) TBI, the neuroimaging tools currently in use are not sensitive enough to detect the locus of damage. Although diffusion tensor imaging (DTI) is showing promise as a more sensitive tool, additional research is needed to evaluate its reliability, validity and ultimate utility. Furthermore, although research has identified links between specific injury sites and changes in cognitive functioning, certain sites are not firmly linked to specific emotional and behavioral consequences.

Children with TBI are at increased risk for social failure as they mature into adulthood. TBI in children is associated with poor academic performance,[13] as well as problem behaviors.[14] It has been estimated that 130,000 U.S. children need special education classes because of TBI but that, in fact, only 11 percent of children with TBI are currently enrolled.[15] These estimates mean that 89 percent of such children remain fully "hidden" to their schools or are misidentified as having other types of emotional or learning disorders.

Thus TBI places a heavy burden on the injured child and adult, as well as on the family. Additionally, in The Incidence and Economic Burden of Injuries in the United States, researchers Eric Finkelstein, Phaedra Corso and Ted Miller estimate the costs to society at $60 billion annually.[16] Because this estimate does not include the costs associated with unidentified TBI, the real figure is higher.

Responding to Traumatic Brain Injury

The key to reducing the personal and societal burdens is to address the needs of people with TBI appropriately. This goal cannot be achieved, however, if people with unidentified injuries remain hidden to themselves and to those who may be helpful in addressing their challenges. As with most health problems, early identification can be lifesaving. Imagine a life, like John's, in which nothing made sense to him or to his family no matter where they turned. Without an explanation, problems are likely to get worse, as a sufferer becomes bewildered and experiences the emotional burden of becoming a stranger to himself and others. Further, parents and teachers often don't link problems in school to an earlier injury and begin to misapply labels that don't help and are likely to do harm.

So, how can "hidden" TBI be pulled into the light and recognized for what it is? First, parents and the medical and educational professionals who address children's needs must become more aware of the potential results of a blow to the head, which any child may experience. Similarly, in the military, officers, medical personnel and loved ones at home must be aware of the necessity to track, for many months, soldiers who have been in combat or have experienced concussive explosions or military accidents, to determine if cognitive, physical, emotional and/or behavioral problems emerge. Such tracking is especially important when soldiers return home and try to pick up the fabric of their former lives, leaving the structure of military life behind them. And last but not least, whenever a concussion or similar injury to the brain is observed or suspected, doctors, family members and friends should take it seriously. In children, surveillance needs to persist over many years, as some problems do not emerge immediately. Unlike adults, children may "grow into" the injury—as they age, their injured brains become unequal to the more sophisticated learning challenges of later childhood, adolescence and adulthood.

In addition, we must develop a good way to screen for brain injury. Community agencies, health-care service providers and other organizations should screen populations that are known to be at risk, such as schoolchildren, abused women, athletes, people receiving social support services and members of the military. Similarly, within medical contexts, people who have experienced a non-brain physical trauma, such as a fall from a ladder, also should undergo screening as a precaution.

Such screening would explain the circumstances that underlie problems and facilitate appropriate diagnosis, possible treatments and accommodations. The Brain Injury Screening Questionnaire (BISQ), which was developed at Mount Sinai School of Medicine, incorporates elements of symptom checklists developed by Donald Lehmkuhl[17] (at The Institute for Rehabilitation and Research) and by the Medical College of Virginia,[18] and is based on the structure of a brief screening tool developed by Meryl Picard, David Scarisbrick and Robert Paluck in 1991.[19] It has been used to conduct brain injury screening in a variety of populations and is the only instrument validated by the Centers for Disease Control and Prevention to screen for a history of TBI.

The BISQ first determines whether minimal criteria for brain injury, as established by the American Congress of Rehabilitation Medicine, are met. These criteria include a blow to the head, a loss of consciousness or a period of being dazed and confused.[20] The questionnaire is particularly effective because, to help jog the memory, particularly about long-ago events, it lists situations in which a brain injury may have happened. If a possible brain injury is so documented, the questionnaire then reviews symptoms that may be present and how frequently they occur. The BISQ can be administered via self-report or can be completed by a proxy. It is now being adapted for administration online, with immediate turnaround of results. The BISQ takes 5 to 20 minutes to complete, depending on whether or not the person has a history of blows to the head. It can easily be administered by social service agencies, in schools, in medical contexts and among at-risk populations.

A point to be emphasized is that the BISQ cannot determine that a TBI is the known source of an individual's problems. BISQ data are used only to generate a statement that a weak, moderate or strong possibility of a brain injury exists. This feedback is based on research showing that 25 of the symptoms in the BISQ checklist are sensitive and specific to TBI—especially cognitive symptoms.[21] People showing more of these symptoms are more likely to have experienced a TBI.

If a person screens positive, three avenues of response present themselves: testing, treatment and accommodations. Neuropsychological testing can more precisely document the nature of deficits that may have been caused by a TBI, and advances in brain imaging could one day be useful as well.

If testing indicates that a TBI has or is likely to have occurred, the next step is to seek treatment or implement accommodations. While available standard treatment typically focuses on cognitive rehabilitation, studies currently under way at Mount Sinai (and elsewhere) focus on treating behavioral and emotional consequences of TBI. For example, we are currently evaluating the effectiveness of a version of cognitive therapy, adapted to accommodate the cognitive challenges of people with TBI, on alleviating post-TBI depression and/or anxiety, and we are examining the efficacy of treatments for executive dysfunction. Accommodating people with TBI is important outside treatment programs, in people's daily lives. Although TBI cannot be cured, families and educators can address its consequences in a variety of ways. Schools should obtain professional input and technical assistance to learn how to accommodate students with TBI in the classroom. Taking steps that permit a student with TBI to learn and to prosper usually helps the other students in the class as well; good teaching for the one is good teaching for all. Individuals who are not in school should seek help from a professional who has experience in dealing with TBI. Such expertise is not available in all areas of the country, but the resources that are available can be found through national and state brain injury associations.

A colleague tells the story of having sat through many meetings where professionals were talking about people with traumatic brain injury and the challenges they face. One day she realized that the cognitive problems she had experienced since childhood may have resulted from the two blows to the head she had experienced as a child as a result of falls; in one incident she remembered, she was lying on the ground in her backyard as a young child, having just fallen from her perch on the top of a swing set, landing on the top of her head and feeling dazed for a while. She had never linked the learning and memory problems she was troubled by over the years to these long-ago episodes. There is no way of knowing without doubt that her continuing problems are the result of the two childhood accidents, but she now conceptualizes her problems in this way, which allows her to explain her challenges and see more clearly the need to compensate for problem areas. For my colleague, for John, for returning soldiers and for the millions of civilians suffering from traumatic brain injury, we must improve our response to this underestimated problem.

CHAPTER 7

Connectomics

Tracing the Wires of the Brain

By Sebastian Seung, Ph.D.

Sebastian Seung, Ph.D., is Professor of Computational Neuroscience at the Massachusetts Institute of Technology and Investigator at the Howard Hughes Medical Institute. He has been a Packard Fellow, Sloan Fellow, and McKnight Scholar. Dr. Seung studies neural networks using mathematical models, computer algorithms, and circuits of biological neurons in vitro.

Scientists working with rapidly advancing computer technology and electron microscopes hope one day to map the billions of neuronal connections in the brain. The resulting map, or "connectome," could help us understand memory, intelligence and mental disorders, Dr. Sebastian Seung writes.

SUPPOSE THAT SOMEONE GAVE YOU A RADIO and asked you to figure out how it works. You could try measuring electrical signals inside it, but the measurements might not be sufficient. You might be more successful if you were also given a circuit diagram illustrating all the components of the radio and how they are connected to each other.

Now imagine that your goal is to discover how a brain works. A map of brain connections would be helpful for interpreting measurements of the signals transmitted between neurons. In the human brain, these signals travel in a complex network of 100 billion or so neurons, each of which is connected to 10,000 others.

Such a map of a brain, human or otherwise, does not yet exist. But as technology advances, researchers are setting their sights on the "connectome," a word coined in a 2005 study by Olaf Sporns and colleagues to describe a complete map of connections in a brain or a piece of a brain.

Genome and Connectome

"Connectome" was coined in analogy with the "genome"—the entirety of an organism's hereditary information—studied by biologists. To imagine how the story of the connectome will unfold over the next few decades, it's helpful to recall the history of the genome.

In 1953, James Watson and Francis Crick proposed the double helix structure for DNA. The double helix consists of a long chain of repeated units called nucleotides, of which there are four types: A, C, G and T. Hereditary information is written in DNA using this alphabet of four letters. In the human genome, the sequence of nucleotides is about one billion letters long. The reading of this sequence was finally completed by the Human Genome Project in 2003.

The story of the connectome began when scientists first realized that the brain comprises a network of neurons. This happened around 1900, well before the double helix of Watson and Crick. But the connectome story is still in the future, and I believe the discoveries that compose this saga will be among the great prizes of 21st-century neuroscience.

Revealing connectomes will be much more difficult than identifying genomes. But we are now optimistic that the connectome will eventually be transformed from dream into reality. A new field of neuroscience will be created: "connectomics." This new field will be driven by new technologies, as we will see. It will take shape alongside other research approaches, and these multiple methods will provide better insight into the brain's complex structure than any individual method can.

Three-Dimensional Nanoscale Imaging

Connectomics is more challenging than genomics; the structure of the brain is extraordinarily complex. You have probably seen images of neurons before. A single neuron has a fantastic shape, forking out many branches to form a tree-like structure. But if you have seen only pictures of neurons in isolation, you may not fully appreciate the complexity of brain structure.

Before researchers study a single neuron under a microscope, they inject it with a stain. The neurons around it remain invisible because without the stain they are transparent. This technique is valuable for seeing the shape of a single neuron clearly. However, it does not give an accurate impression of what the brain is really like because neurons are not islands in the brain. Instead, their forking branches are tightly entangled with each other. The brain can be compared to a giant bowl of spaghetti, in which each strand has been replaced by a complex, branched noodle.

Because their branches are so tightly entangled, neurons are locked in a multi-way embrace. At a point of contact between a pair of neurons, they can form a synapse, a junction at which one neuron sends chemical messages to another. When a synapse exists, the pair of neurons is said to be "connected." The term should not be taken too literally, as there is still a narrow gap separating the two neurons, and the molecules in chemical messages have to float across this gap. The term is used in the metaphorical sense of communication, just as two people talking on cell phones are said to be connected. The efficacy of communication between a pair of connected neurons is known as the "strength" of the connection. If two neurons are strongly connected, the messages between them come in loud and clear, but if they are weakly connected, the messages are faint. Entanglement increases contact points between neurons, providing more potential locations for synapses, which allow neurons to communicate, or be "connected."

Although entanglement is a crucial aspect of brain structure, it's impossible to see with an ordinary light microscope. According to the laws of physics, structures smaller than the wavelength of light cannot be seen clearly using such a microscope.* The thinnest branches of neurons are less than a tenth of a micron in diameter,† which is less than the wavelength of visible light. Luckily, another kind of microscope uses electrons rather than light and yields images with much higher spatial resolution. With an electron microscope, the branches of neurons can be seen clearly, even when they are tightly packed together in the brain.

By itself, a microscope cannot be used to see the interior of the brain, which is essential for observing brain structure. So, to see every location in the brain, scientists slice brain tissue into thin sections with a knife. By the combined use of the knife and the microscope, a sequence of two-dimensional images is acquired. Together these images show the entire three-dimensional brain structure.

---

* In the field of optics, this limitation is known as the Rayleigh criterion. Only recently, researchers in the emerging field of nanoscopy have figured out that there is a loophole, so that fluorescence microscopy can beat the Rayleigh limit. This discovery suggests the exciting possibility that connectomics could eventually make use of light microscopy as well as electron microscopy.

† There are 1,000 microns in a millimeter, so 10,000 of these thin branches could be laid side by side into a millimeter.

Image Analysis

The images from three-dimensional electron microscopy are generally visualized in the same way they were acquired, as a temporal sequence of two-dimensional images—in other words, a movie. A video available online* contains images from the laboratory of Dr. Winfried Denk, a German scientist working at the Max Planck Institute in Heidelberg. The images show the structure of the retina of a rabbit. The retina, located at the back of the eye, is a sheet of neural tissue that converts light into neural signals.

Each frame of the video shows one slice through the retina. As the video plays, you are looking deeper and deeper into the specimen of neural tissue. The neural tissue consists of many branches of neurons, tightly packed together. Therefore, a cross section through the neural tissue reveals the cross sections of numerous branches, each a small white region surrounded by a dark contour.

In addition to the gray-scale images, notice the colored objects in the video. They are branches of neurons—the "wires" of the brain—traced through the images by computer. You can also inspect the neural branches in the accompanying images, which are still images from the video.

In principle, humans can do this tracing manually. In practice, doing so would be painstaking and laborious, as the branches of just a single neuron could take many hours to trace. Furthermore, even a small sample of neural tissue comprises many branches. It would take an incredible amount of human effort to manually trace all of the branches in just a cubic millimeter of neural tissue. So we are turning to computers to automate the process of tracing and reduce the amount of human effort required.

This research to study neural wiring is in its beginning stages, so the algorithms are still not very good. The computer makes many mistakes. Sometimes it merges two neurons into one object. Sometimes it splits one neuron into two. As in the evolving field of computer vision, a branch of

---

*Visit http://hebb.mit.edu/people/viren/videos_sfn/
sfn_640480_maintitles_h264_medium.mov (accessed January 29, 2009).

artificial intelligence in which robots try to see and recognize the objects around them but still cannot reliably do so, we do not yet have computers that can accurately see the shapes of neurons. Unless we can make such computers, we cannot hope to replace human effort with computer automation to trace the shapes of neurons.

Nevertheless, I am optimistic that researchers will eventually solve these technical problems. Then we will be ready to find connectomes. We will start with the simplest of brains, or even small parts of brains. As our technical abilities develop, we will scale up to larger brains. The process will start with electron microscopy, which will generate three-dimensional images of a specimen of brain tissue. Each synapse will be identified, and the branches involved will be traced back to their parent neurons. By assigning each synapse to a pair of neurons, we will be able to map the connections in the brain. The connectome will begin to take shape.

Today a small community of scientists and engineers is working on developing these technologies. If they are successful, the new field of connectomics will change our understanding of the brain in myriad ways.

The Connection Theory of Memory

First, connectomics will help reveal how the brain stores and retrieves information about the past. Neuroscientists believe that memories are stored in the connections between neurons. According to this theory, connections change when a new memory is stored. That such changes can happen is not in doubt. Neuroscientists have found that new synapses can be created and that the strengths of existing synapses can be altered. What remains uncertain is whether these changes are indeed the basis of memory.

Although the connection theory of memory is widely believed, it has been difficult to test experimentally. One barrier has been the lack of good techniques for measuring whether two neurons are connected and, if so, how strongly. One important task of connectomics will be to determine the connectivity of brain areas that are involved in memory storage.

Take, for example, sequential memory, such as the notes of a piece played on the piano. A pianist is able to store such sequences in memory and recall them at will. Recently, sequential memory has been studied by neuroscientists in the brain of the zebra finch. This bird learns a single, highly repetitive song as a juvenile and sings it repeatedly as an adult. The avian brain area called HVC appears to be important for a bird's memory of its song. Creating a lesion in the HVC in adult birds causes a loss of song. Furthermore, electrical recordings of "projection" neurons in the HVC, which send long branches (axons) to downstream brain areas, have revealed a precisely timed, repeated sequence of neural activations while the bird sings this song. The zebra finch song consists of repetitions of a single motif, which is about one second long. During a song motif, each projection neuron in the HVC is activated exactly once, for just a few milliseconds. For every repetition of a motif, the projection neurons are activated in the same sequence. The projection neurons of the HVC activate other neurons in an area called RA, which in turn activate the motor neurons that control the syrinx, the avian vocal organ. Therefore the sequential activation of HVC neurons is believed to be directly responsible for the sequence of movements that produces birdsong.

According to one theory, the memory of the sequence is stored in the connectivity of the HVC. Each projection neuron transmits an excitatory signal through its synapses onto the neurons that are just after it in the sequence. These synaptic connections cause the neurons to be activated in sequence, like a row of falling dominoes.

Currently, it is not feasible to make the measurements that would be required to test this theory about HVC connectivity. But connectomics will eventually make it possible to find out whether the memory of birdsong is indeed stored in a sequential organization of connectivity in HVC. I expect that it will become possible to study the connectional basis of other types of memory as well.

The Connection Theory of Intelligence

Connectomics also holds the potential for addressing the biological bases of mental—cognitive and emotional—differences. Just as physical characteristics differ, so do mental characteristics. For instance, some people get angry easily, while others seem to show little emotion. The few who have prodigious capabilities, such as Mozart and Einstein, are called geniuses. What causes these mental differences?

In the 19th century, the English scientist Francis Galton proposed that smart people have larger brains. He tested his proposal by comparing the head sizes of students at Cambridge University with their grades. Modern researchers continue in the tradition of Galton but use the more sophisticated method of magnetic resonance imaging to study brain structure. They have confirmed a correlation between overall brain volume and intelligence as measured by IQ tests. However, the correlation is fairly weak.

As brain imaging methods have become more advanced, researchers have been making more precise measurements of brain structure. They are looking for some structural feature that is more strongly correlated with intelligence than is overall brain volume. Some have looked at the sizes of particular brain areas. Others have examined the brain's white matter, the neural fibers that connect brain areas. Still others have looked at the changes in the thickness of the brain's cortex during development.

With the advent of connectomics, it will be possible to investigate a related but somewhat different hypothesis. Maybe intelligence depends not on the size of a brain but rather on how its connections are organized. Because most of the volume of the brain is devoted to the neurons and their synaptic connections, differences in connectivity could be crudely manifested as differences in brain volume. However, there could be many differences in connectivity that have no effect on brain volume and that are difficult or impossible to measure by current techniques.

It has been popular to think of the brain as an assembly of interacting modules, each of which is functionally specialized and localized in a distinct brain area. One could imagine that mental differences arise from differences in connectivity within modules, between modules or both. The connections between modules are like the superhighways of the brain. They are relatively wide and long compared with connections within modules, which are like local streets and alleys. Connectomics will provide improved methods of studying both types of connections. With these improved methods, I think we will be able to find out whether mental differences are related to differences in connectivity.

Even if mental differences are caused by brain differences, that does not mean they are exclusively or even primarily determined by genetics. Some are the results of life experiences rather than heredity. For example, although identical twins have identical genes, one may play the piano and the other may play the flute. Much as muscles can be changed by weight lifting, brains can be changed by practice.

Connectopathies

In addition to studying differences in normal mental functioning, connectomics will address the causes of mental disorders. In some cases, these disorders are characterized by the death of neurons—neurodegenerative disorders such as Alzheimer's disease, for instance.

But for other mental disorders, neuron loss appears far less significant. Rather, many scientists suspect that at least some mental disorders involve abnormalities in the connections of the brain, or "connectopathies." Since the 19th century, when this hypothesis was first proposed, its popularity has waxed and waned more than once. Today it is popular again, as scientists use magnetic resonance imaging to look for abnormalities in neural networks—connectopathies—that may be associated with disorders like schizophrenia and autism. Investigators speculate that such disorders are developmental, perhaps caused by abnormalities in the processes by which the brain wires itself. Imaging research is complemented by progress in finding genes that are associated with mental disorders. Some of these genes could be involved in controlling brain-wiring processes.

Although magnetic resonance imaging has the advantage that it can view live brains, its disadvantage is low spatial resolution. Connectomics will make it possible to map brain wiring at a much higher resolution, thus providing more refined methods of looking for connectopathies.

The Future

These proposed applications of connectomics to the human brain, with its 100 billion neurons and their connections, are very speculative. Finding the human connectome is not realistic now. Connectomics must begin instead with simple brains.

For example, the worm *C. elegans* is about 1 millimeter long. While it doesn't have a brain like ours, it does have a nervous system containing 300 neurons and 7,000 synapses. Every connection of this nervous system was mapped in the 1970s and 1980s by a team of scientists. Because the analysis was done manually, without the aid of a computer, it took more than 10 years to complete.

With the technologies currently under development, the labor involved in slicing, imaging and analyzing the images will be reduced through automation. That will make it possible to find the connectomes of brains more complex than the *C. elegans* nervous system. At first these brains will be very small—the 100,000 or so neurons in the fruit fly brain, for example. Finding even this small bug's connectome is an incredible challenge.

We will also try to determine the connectomes of small parts of large brains. A prime target is the retina, the sheet of neural tissue at the back of the eye that converts visual stimuli into neural signals. Although the retina is in the eye, it contains networks of neurons much like those in the brain and is actually considered part of the brain because of its common embryological origin.

If we are successful in meeting these challenges, then we will move on to the brains of larger animals, such as the mouse or bird. How far will technology progress? Will I live long enough to see the human connectome? Finding the human connectome will be one of the greatest computational challenges of all time because of the difficulty of analyzing such a large amount of image data. A 1-terabyte hard disk costs a few hundred dollars today; you would need 100 *million* terabytes to store the images from a human brain if they were taken at 20-nanometer resolution and not compressed by software. (There are 1 million nanometers in a millimeter.) Even with 10,000 microscopes working in parallel, it might take

30 years to collect all of the images. Analyzing the images could require a parallel supercomputer with millions of processors.

Of course, it's difficult to predict the future, but the rate of technological progress in other areas gives reason for optimism. Computer technology is still improving exponentially over time. The price of computation is halved every two years, and the price of storage is halved every year. Astonishingly, the pace of improvement in genomic technology is even faster than that. The price of finding a human genome has dropped by a factor of 10 every year for the past 4 years.

If computer technology were to improve at an exponential pace for several more decades, then we could be confident about eventually finding the human connectome. Some are predicting that the exponential party will come to an end in just one decade, but others believe that advances in nanoelectronics will keep it going for longer.

Therefore, humanity's quest to understand the brain will go hand in hand with the quest to build ever more powerful "artificial brains"—computers. I am confident in the future of both endeavors. I believe that by the time the 21st century is recorded in the history books, one of its greatest triumphs will be the determination of the intricate and awesome structure that gives rise to the mind: the human connectome.

CHAPTER 8

The Meaning of Psychological Abnormality

By Jerome Kagan, Ph.D.

Jerome Kagan, Ph.D., emeritus professor of psychology at Harvard University, was co-director of the Harvard Mind/Brain/Behavior Interfaculty Initiative. He is a pioneer in the study of cognitive and emotional development during the first decade of life, focusing on the origins of temperament, and is the author or co-author of more than 20 books, including the classic Galen's Prophecy: Temperament in Human Nature (Basic Books, 1994).

Widespread diagnoses of childhood disorders trouble scientists such as Dr. Jerome Kagan, who argues here that social conditions, not biology, are often to blame. Kagan elucidates possible reasons for the increase, citing, among other explanations, pressures on parents to raise flawless children. He concludes by proposing ways to avoid misdiagnoses in the future.

THE RECENT INCREASE in the number of children diagnosed with autism, bipolar illness, or attention-deficit/hyperactivity disorder (ADHD), widely reported in the media, has created worry both among the public and among health officials. It is important, therefore, to ask whether this troubling trend reflects a true rise in mental illness or is the result of changes in the definition of childhood psychiatric disorders. The latter explanation is likely because the concept of psychopathology is ambiguous, and physicians have considerable latitude when they classify a child as mentally ill. Because a diagnosis of ADHD, bipolar disorder, or autism allows parents to obtain special educational and therapeutic resources that would not be forthcoming if the child is called mentally retarded, incorrigible, or uninterested in academic progress, doctors are motivated to please the distraught parents who want to help their child.

Psychiatrists diagnose a mental disorder when a set of behaviors or emotions is infrequent (usually possessed by less than 10 percent of the population); when the child or family is distressed by the symptoms; or when the symptoms interfere with the child's adaptation to his or her society. Often distress and poor adaptation occur together. But because both the frequency of a symptom and its adaptive qualities change with history and across cultures, the prevalence of many mental illnesses also changes. For example, most children living in the American colonies during the 17th century were not required to maintain attention on an intellectual task for 5 or 6 hours a day—and there was no concept of ADHD. Social phobia also was less common because youths knew almost everyone in their village or small town. Although a Puritan child who resisted most parental requests would be classified as deviant and in need of help, psychiatrists today would not have classified a 6-year-old who resisted most parental requests as having oppositional disorder if neither the parents nor the child was concerned with the latter's autonomous behavior and the child performed relatively well in school.

Changing Social Challenges

Each historical era within a society poses special adaptive challenges for its members, and traits that would be regarded as maladaptive and possible signs of a disorder in one era might be more adaptive in another. For example, adolescents who wore sexually provocative clothing and rejected the existence of God would have been both rare and a source of parental worry in 17th-century Massachusetts, but today these traits would not be regarded as signs of pathology.

In addition to changing definitions of pathology, three other factors are contributing to the apparent epidemic of childhood psychopathology in economically developed societies.

First, the contemporary American economy requires every child to complete high school with adequate language and mathematical skills, and preferably go on to receive a college degree, in order to obtain a job with some financial security. These were not requirements in the 18th century; Benjamin Franklin did not have the advantage of a high school education. The pragmatic requirement for academic accomplishment generates worry in parents who are concerned about their child's future. Hence, they become anxious if they believe their preschool child shows signs of future academic difficulty.

Second, the availability of technologies that detect serious biological problems in the unborn child enables parents to make the legal choice of aborting these embryos. Because many do so, the proportion of severely abnormal children today is smaller than it was a century ago. Hence, a child with an obvious abnormality, such as Down syndrome, has become more conspicuous, leaving parents more vulnerable to a blend of anxiety, shame and guilt if they sire a child with observable symptoms that are stigmatizing. It is not surprising, therefore, that many parents, especially those not considering abortion, are eager to take advantage of diagnostic techniques that might reveal a potential problem; if there is one, therapeutic interventions can then be implemented early in development.

The American ethic of egalitarianism, which obligates each individual to award dignity and respect to all citizens independent of their values or practices, is a third factor contributing to the increase in diagnoses with a genetic cause. This moral imperative makes it more difficult to blame parental neglect or ineffective socialization practices as contributors to aggressive behavior or poor academic performance and easy to award power to genes for which no one is responsible. Such an attitude frees parents of excessive guilt for the undesired symptoms and protects them from community criticism. The availability of technologies that assess genomes, along with the media's advocacy of biological determinism, has persuaded many Americans that genes must be exceedingly potent, even though no scientist has found any particular gene, or cluster of genes, that is a consistent correlate of poor attention skills, hyperactivity, aggressive behavior, academic failure, chronic disobedience or excessive shyness, independent of the child's social class, ethnicity, cultural background, gender and history of experiences.[1]

The Diversity among Current Diagnostic Categories

Most diagnoses of child pathology are based primarily on symptoms, usually described by a parent, rather than on a combination of these symptoms, the child's behavior on a set of psychological tests, and physiological markers that might represent a foundation of the illness. Although most psychiatric disorders are currently defined as if each were one illness, most have multiple origins in a unique profile of genes and personal history. However, a majority of American psychiatrists place greater emphasis on the genetic roots than on the experiential evidence. This bias has the serious disadvantage of predisposing psychiatrists and pediatricians to prescribe drugs as the only therapy, even though non-drug therapies that take experiences into account often reduce the severity of symptoms.

More than 90 percent of childhood diagnoses are based only on the parents' descriptions of the child without the addition of psychological testing and biological measures, which would provide a more accurate basis for diagnosis. For example, some children diagnosed with ADHD have difficulty maintaining prolonged attention to others' speech but attend well when playing soccer on the playground and are neither hyperactive nor excessively restless. Other children are restless in the classroom but have no problem paying attention. Only a small proportion display both symptoms. Nonetheless, many pediatricians classify all three types of children as having ADHD and prescribe the drug Ritalin because, as I noted, parents can request special educational advantages for children with this diagnosis and the physician would like to satisfy the parents.

Equally serious, if not more so, is the dramatic rise (more than 40 percent in the past decade) in diagnoses of bipolar disorder in young children, based on parental complaints of chronic levels of extreme disobedience, impulsive bursts of aggression, and an inability to control emotion. These symptoms may in part be the product of permissive socialization practices by parents who are reluctant to induce anxiety or guilt in children placed in surrogate care because both parents are working. Most children classified as bipolar do not display the cycles of manic excitement and depression that define this disease in adults. Thus, it is a diagnostic error to call children who cannot regulate their moods "bipolar" simply because they seem to have a single feature in common with the adult disorder: uncontrolled behavior. I do not believe that psychiatrists have detected a new childhood disorder; they have used a new term for a serious rise in poor regulation of emotion that is probably a result of experiential rather than genetic factors.

One reason the diagnosis of bipolar disorder in young children has become popular is that the adult form has a genetic component. Hence, physicians would like to assume that the child's symptoms are also the product of genes, rather than a combination of a temperamental bias, which interferes with effective regulation of behavior, and family practices. Hence, drugs are the preferred therapy, rather than suggestions to parents as to how they might behave in order to alter their child's behavior.

Diagnoses of autism are also on the rise. The symptoms that lead to this diagnosis are serious impairment of cognitive functions, especially language, compromised social skills, inappropriate emotion and stereotyped motor acts, such as pulling at the hair, rocking and head banging. These features, alone or in combination, can be the product of a large but still unknown number of distinct biological conditions, including altered genes or chromosomes, maternal illness during pregnancy, early postnatal infections or rare immune reactions in the young infant.[2] When I was a graduate student in the 1950s, most of these children were assigned the amorphous label "brain damaged." The word "damaged" has stigmatic connotations that are missing from the category "autistic."

Parents of autistic children have persuaded the federal government and private foundations to increase funding for research that in theory will find the cause of what is actually a number of different conditions; the habit of diagnosing a child as belonging to an "autistic spectrum" implies, incorrectly, a single cause for a condition that varies in severity. This is not a wise strategy, and applying it to other disorders sounds absurd. No physician would place a child with a headache in a category called "headache spectrum" because scientists know many of the reasons why a child might have a headache—a brain tumor, concussion, infection, high blood pressure, among others. Similarly, oncologists do not use the concept "cancer spectrum" because they have learned that, for example, leukemia, tumors of the colon and breast cancer are caused by different genes and life experiences. The odds of finding the gene that is a risk factor for one form of breast cancer would be very low if investigators lumped all patients with any form of cancer into one category called the "cancer spectrum."

Because we do not know the host of biological conditions that can cause the symptoms that define autism, the concept of an autistic spectrum survives and is interfering with research attempts to find the many separate causes of these traits. The habit of pooling into one group all the children given the diagnosis of autistic spectrum makes it difficult to discover a particular etiology that affects a small proportion with some of the unique symptoms that define this family of illnesses.

New Strategies

The adoption of two practices will have benevolent consequences. First, every child who appears to have a psychiatric illness should have psychological tests administered to determine, with greater specificity, the exact pattern of cognitive and emotional impairments, as well as appropriate assessments of brain and bodily function. An example: among children who are excessively shy and timid, those who also display greater neuronal arousal in the frontal lobe of the right hemisphere, rather than the left, along with greater activation of a cluster of neurons in the brain stem that respond to sound are most likely to remain shy and timid through adolescence.[3] These two distinct biological features can be measured using an electroencephalogram, which detects changes in brain activity via painless electrodes placed on the scalp. This additional evidence would allow investigators and clinicians to parse the currently heterogeneous category of social anxiety disorder into those who do and those who do not possess the two biological features. The therapeutic recommendations should be different for these two forms of social anxiety.

Second, scientists and clinicians should not equate the presence of an abnormal gene with the existence of a disease. We don't make such a jump in other circumstances. For example, most adults with Helicobacter pylori in the stomach do not have ulcers; most with staphylococci bacteria on their skin do not develop an infection. I suspect that at least half of the children currently diagnosed with ADHD, who take Ritalin in order to increase dopamine activity in the brain, do not have a genetic compromise in dopamine function. A careful examination of the child and the home environment might reveal why a particular child is having difficulty paying attention in school.

It is important to appreciate that the majority of children under age 6 with some form of psychopathology will not become seriously disturbed as adults, especially if they are growing up with nurturing, middle-class families, because most children grow toward health.[4] My colleagues and I are studying a large sample of children observed first at 4 months and evaluated regularly through early adolescence. Most of the extremely shy, fearful 2-year-olds who were temperamentally vulnerable to anxiety over novel events (about 20 percent of the sample) were well adapted to their school and social settings by the time they were 15 years old. Only about one-third of this group were still socially anxious and unhappy with their personality.[3] Scientists who study conduct disorder report that the majority of middle-class children under age 10 who are diagnosed with oppositional or conduct disorder do not become criminals or delinquents.

Critically, the social class in which the child is reared is far more predictive of future psychopathology than is a particular gene or type of behavior displayed at two years of age. Anxiety disorder, depression, criminality and drug addiction are more frequent among youth and adults who grew up in poorer families with less educated parents, whether in developed or less well developed societies.[5] The fact that academic failure, conduct disorder, substance abuse, anxiety disorder and depression are more strongly associated with the education and income of the family of rearing than with a particular gene implies that scientists have to balance their study of the genetic causes of illness with probes of the environmental contributions, and clinicians should consider therapeutic strategies other than drugs. At the moment, public and private agencies allocate more research funds for study of the biological rather than the psychological circumstances that underlie disorders. This imbalance is odd in a society that prides itself on being practical and hostile to abstract philosophical arguments. The American attraction to material causes, and the presumed certainty that they promise, sustain the belief that biological therapeutic interventions will be more effective and easier to implement than sociological or psychological ones. This premise has defeated much of the motivation to alter the life conditions of economically compromised members of our society and to improve the quality of the schools they attend.

Bouts of anxiety, fear, guilt, shame, sadness, depression, anger and envy are inherent in the human condition because, after age 2, humans can anticipate the distant future, possess a moral sense, and wish to regard the self as worthy and good. A few children experience the above disturbing emotional states far more often than their peers do. But the sharp rise in the incidence of bipolar disorder in children is more likely the result of changing social conditions, parental practices and/or diagnostic criteria than the discovery of a new genetically mediated illness. The current confusion among parents over the best way to raise children and the ethical values to instill, the large gap in family income between the top and bottom quartiles that has developed during the past 25 years, and the need to accommodate to the extraordinary diversity in the values and socialization practices of the many ethnic groups in American society have generated parental interactions with children that have produced profiles that fit the criteria of a mental disorder. Yet, surprisingly, most scientists are searching only for the genetic contribution to these symptoms. This bias is not new; the 19th-century French attributed the presence of large numbers of prostitutes to their degenerate heredity rather than their life of poverty.

I am not claiming that all psychopathology is socially mediated. There are individuals with hallucinations, delusions, panic attacks, manic moods, crippling levels of anxiety and numbing depression who inherited a biological disposition that rendered them vulnerable to these debilitating symptoms. However, the prevalence of these serious pathologies has remained relatively stable over long periods of time. If there has been a real rise in one or more of the categories of childhood pathology, it is likely that changing social conditions and altered diagnostic criteria are the major reasons for this disturbing phenomenon.

CHAPTER 9

The Impact of Modern Neuroscience on Treatment of Parolees

Ethical Considerations in Using Pharmacology to Prevent Addiction Relapse

By Richard J. Bonnie, J.D.,
Donna T. Chen, M.D., M.P.H,
and Charles P. O'Brien, M.D., Ph.D.

Richard J. Bonnie, J.D., is Harrison Foundation professor of medicine and law, professor of psychiatry and neurobehavioral sciences and director of the Institute of Law, Psychiatry and Public Policy at the University of Virginia. He specializes in mental health and drug law, public health law and bioethics.

Donna T. Chen, M.D., M.P.H., is an assistant professor in the Department of Public Health Sciences and the Department of Psychiatric Medicine and a member of the Center for Biomedical Ethics at the University of Virginia. She has served on a number of national committees devoted to clinical and research ethics.

Charles P. O'Brien, M.D., Ph.D., is professor and vice chair of psychiatry, University of Pennsylvania, and chief of psychiatry, director of the Center for Studies of Addiction. His research focuses on the mechanisms of addiction and possible pharmacological treatment of related disorders.

The authors have no pharmaceutical stocks or patents. In the past three years, Charles P. O'Brien has served as a short-term, paid consultant to The Center for Victims of Torture, Alkermes Inc., Eli Lilly and Company, Merck and Co., Inc., Purdue Pharma LP, Ortho-McNeil-Janssen Pharmaceuticals, Inc., Cephalon Inc. and Pfizer Inc.

As neuroscience advances our understanding of addiction, a drug called naltrexone offers the possibility of treating drug offenders, particularly those on probation or parole, and helping them avoid relapse. The Dana Foundation supported a pilot feasibility study on this approach, conducted by Charles P. O'Brien and colleagues. Here, Richard J. Bonnie, Donna T. Chen and O'Brien examine ethical and legal concerns related to various methods of administering naltrexone.

UNTIL RECENTLY, the nation's policymakers have seemed indifferent to the high costs of criminal punishment of drug-addicted individuals and the apparent failure of criminalization to reduce heavy drug use. Increasing frustration with the costs of criminalization has fueled renewed interest in treatment of addicted offenders and others with drug problems, especially at the state level. We see a major opportunity for the criminal justice system to take advantage of recent advances in the pharmacological treatment of opioid addiction—in particular by facilitating the use of an injectable drug called naltrexone to prevent relapse by individuals who are under community legal supervision (probation or parole). This initiative is an attractive first step because most of these individuals should be highly motivated to stay "clean" in order to stay out of prison, society has a strong interest in reducing addiction-related crime, and the extended-release formulation of naltrexone has been approved by the Food and Drug Administration for the treatment of alcoholism and soon may be approved for treatment of opioid addiction as well. But using naltrexone to prevent relapse and repeat offenses does raise ethical and legal concerns.

Background

The costs of criminal punishment for drug-addicted individuals are borne by these individuals, their families and friends, and society—members of the community who fall victim to drug-related crimes and others whose tax dollars pay for incarceration, a strategy that has repeatedly been shown to be ineffective in reducing personal and social problems caused by drug addiction. A 2001 report issued by the National Research Council of the National Academy of Sciences calls for more research on the effects of criminal sanctions against users, characterizing the government's apparent indifference to the effects of its costly policies as "unconscionable."[1]

A renewed interest in a therapeutic response to addicted offenders is reflected in the creation of specialized drug courts (numbering 2,147 at the end of 2007), legislation requiring treatment for nonviolent drug-involved offenders, and other programs.[2] Surprisingly, however, this therapeutic movement has not taken full advantage of neuroscience advances in the understanding of addiction and associated pharmacological developments.

The drug treatment offered in the criminal justice system is largely educational and counseling-oriented. Few programs offer medications, even ones with significant evidence of effectiveness in reducing drug use and criminal behavior. Legislators and judges seem to us to be deeply skeptical of methadone maintenance for opioid addiction (believing that it only substitutes one addictive drug for another) and have generally overlooked other medications. These include buprenorphine (like methadone, this long-acting medication activates the opioid receptor, and it has been shown to decrease heroin craving) and naltrexone (a medication that blocks the effects of opioids at the receptor and thus blocks the high from heroin and other opioids). The gap between therapeutic opportunity and actual clinical practice in addiction treatment (not only in criminal justice populations) is about to widen with the introduction of extended-release versions of existing drugs and rapid advances now being made in the development of other anti-craving drugs and, potentially, vaccines targeted at drugs of abuse. In short, we appear to be on the threshold of major advances in the pharmacological management of addiction.

In choosing to focus on extended-release naltrexone in the setting of community legal supervision, we set aside many other ways that neuro-scientific advances in understanding and treating behavioral problems such as addiction might affect criminal justice policy. For example, we do not engage the larger question of the appropriateness of criminalizing drug use by addicted offenders, although this broad issue is being considered by a new project on law and neuroscience supported by the John D. and Catherine T. MacArthur Foundation.[3] Rather, we focus narrowly on ethical, legal and practical aspects of how the criminal justice system might manage offenders who are addicted to heroin or other opioids by facilitating use of extended-release naltrexone.

Naltrexone for Prevention of Relapse in Parolees

Heroin-addicted individuals commit many crimes to support their habits. Although these are usually nonviolent crimes, addicted offenders are typically sentenced to prison terms. Despite not using opioids for a long period, incarcerated opioid addicts relapse at an alarming rate following their release from custody, even when they are under the supervision of a parole officer. It may be thought that the period of incarceration would "get the opioids out of their system" and "teach them a lesson," but this is apparently not enough to prevent re-addiction and re-incarceration in a majority of opioid-addicted offenders. The availability of naltrexone may provide real benefits to these individuals, the criminal justice system and the public at large.

Studies conducted in the 1990s suggested that daily ingestion of a naltrexone pill taken by mouth reduced the frequency of relapse in parolees. Naltrexone has a high affinity for opiate receptors and prevents the high from heroin or other opioids by blocking their access to the receptor. In the only randomized, controlled clinical trial of probationers with a history of opioid addiction, Dr. James Cornish and colleagues in Philadelphia found that 59 percent of opioid-addicted parolees who received standard parole supervision—but not naltrexone—relapsed and were re-incarcerated within a year of their release. In contrast, a randomly assigned group of similar parolees who received both standard parole supervision and naltrexone from a research nurse stationed at the parole office had a relapse rate of only 25 percent.[4]

Notably, treatment with oral naltrexone was approved by the FDA in 1985 and is not considered experimental. It does not require stringent ethics reviews and heightened informed consent procedures any more than does the use of any other FDA-approved medication. It is not addicting, has few side effects and is not considered to be a dangerous drug. It does have a "black box" warning in the package insert about the potential for liver damage, but this warning was based on patients in the 1970s who received seven times the recommended dose in an effort to reduce appetite. Clinicians consider it to be a safe medication.

Adherence and the New Naltrexone Formulations

One of the problems associated with oral naltrexone is the low rate of adherence to daily ingestion. Patients forget to take the pill or resist taking it—thus limiting its potential effectiveness. Beyond these general factors, which are common to most medications and patient populations, special circumstances limit naltrexone adherence. First, many opioid-addicted patients have disorganized lives, which can lead to forgetting. Second, naltrexone prevents the euphoria from heroin or other opioids—which, consciously or unconsciously, the individual may not wish to give up. Finally, there are no reinforcing pharmacological effects from taking naltrexone as one gets from taking methadone or buprenorphine—it doesn't make the patient feel good.

Various companies have been working to develop sustained-release versions of naltrexone ("depot naltrexone") to reduce adherence problems. A single injection given every 30 days provides continuous, steady-state medication throughout this time period. The extended-release form of naltrexone was approved by the FDA in 2006 for the treatment of alcoholism and is expected to be approved for the prevention of relapse to opioid addiction in the near future.

The National Institute on Drug Abuse (NIDA) has recently funded a five-site study of recently released parolees with a history of opioid addiction to be treated with either extended-release naltrexone or treatment as usual for six months. This study will be completely voluntary, as described later. Beginning the treatment in the community is a good first step. However, we believe that the most efficient and effective way to use this medication would be to initiate the treatment prior to an addicted individual's release from prison—in pilot studies, most of those volunteering for treatment after release had already relapsed and had to be detoxified before starting naltrexone.

Options for Criminal Justice Policy

Assuming that the FDA approves the use of extended-release naltrexone for opioid-addiction relapse prevention, there are three approaches that criminal justice policymakers could take toward facilitating its use:[5] (1) a "voluntary" approach, in which the treatment is not linked to the offender's status in the criminal justice system and the offender's decision to participate (or not) in treatment and to take (or not) naltrexone is unequivocally voluntary; (2) a "leveraged" approach, in which the offender agrees to undertake the treatment in return for a more favorable disposition of the case; and (3) a "no choice" approach, in which the offender is simply ordered by the court to take the drug. Each of these approaches raises distinct ethical concerns.

The Voluntary Approach

The voluntary approach does not link participation in drug treatment to an individual's criminal justice status. An offender on probation or parole is legally obligated to remain clean and may be subject to periodic urine testing. However, unless the court or parole agency specifically makes treatment a condition of probation or parole, the offender is not legally required to participate in treatment, much less to take naltrexone, and he or she may choose not to do so.

The voluntary option is theoretically appealing because it fits into fundamental notions of ethical health care and basic human rights. It recognizes that most offenders should have a strong motivation to stay clean, knowing that relapse will lead to re-incarceration. And, if sufficiently motivated, they will seek out drug treatment, including naltrexone, on their own. Experience has shown, however, that not all drug offenders have, or can sustain, this type of internal motivation, and even when they do, internal motivation is often not enough when it comes to sticking with treatment and staying clean. The underlying problem is that the desire for the high in the short term can all too often overwhelm the addicted person's genuine wish to make himself or herself better off in the long run.

Research shows that removing impediments to treatment and even offering incentives are associated with higher rates of treatment participation. In the context of voluntary use of naltrexone by addicted offenders, the long-acting preparation is extremely promising because of convenience: staying clean would not require daily decisions to stay with the program in the face of temptation. However, even if an offender is psychologically committed, cost may be a barrier. Consider the challenge faced by indigent offenders and their families if they seek out treatment upon release from prison and find that it will cost them $500 to $700 per month in out-of-pocket expenses because they have no insurance. Even if they are fortunate enough to have insurance, lack of coverage for naltrexone or unaffordable insurance copays could very well make the drug financially inaccessible.

Because most individuals in the criminal justice system likely will not have the resources to obtain naltrexone on their own, particularly early in its patent life, a significant policy matter—and indeed a significant ethical matter—is whether such treatment should be subsidized for any addicted offender who wants it. The NIDA-funded study mentioned earlier, which is using extended-release naltrexone with offenders who are under community supervision and who have volunteered to participate, will help inform policymakers about the effectiveness of this policy option in reducing relapse and repeat offenses.

It is not clear how many drug offenders would take advantage of naltrexone under a purely voluntary approach—at least in the absence of a major commitment by the criminal justice system to subsidize it. Some will decline because they believe they can beat the addiction on their own, others because they are unable to harness the psychological commitment needed to become and/or stay clean. Here, too, research shows that when people are able to prod themselves into initiating treatment, they often have difficulty sustaining participation on their own. But external pressure, ranging from informal nudging on the part of a spouse, to more formal interventions by employers, to legally mandated treatment through a court order, can help people do what they know is in their long-term interest even if they are unable to do it in the face of strong short-term desires. Just think of all the strongly motivated people who relapse while trying to quit smoking, even after being clean for months. Studies suggest that individuals pressured into treatment do as well as those who seek it voluntarily, and some develop the internal motivation to remain in treatment and/or stay clean after initially being pressured into treatment.

The Leveraged Approach

Leveraged approaches to inducing acceptance of drug treatment already exist. A leveraged agreement connects a person's legal status in the criminal justice system with participation in drug treatment, typically through plea bargaining for a probationary sentence in the community instead of incarceration (or, for someone who is already incarcerated, through agreement to participate in return for early release on parole). Variations on this basic legal arrangement typically used by drug courts include participation in treatment in return for eventual dismissal of the criminal charge or the expungement of a guilty plea. Under such a leveraged agreement, a person might agree to participate in treatment with or without agreeing specifically to take naltrexone.

Leveraged participation in drug treatment generally includes frequent monitoring of compliance with the treatment program's requirements (e.g., attendance at counseling sessions) and of drug use (e.g., clean urine samples), as well as use of incentives to reward adherence and sanctions to penalize undesired behaviors. At issue, then, is whether extended-release naltrexone could be part of such a treatment program.

Some civil libertarians and treatment providers oppose leveraged use of extended-release naltrexone because they regard leveraged agreements in the criminal justice system to be inherently coercive. We do not agree with this point of view. Rather, we believe that offering drug-addicted individuals treatment instead of incarceration expands their options and that acceptance of treatment, though leveraged, is not coerced. Similarly, we believe that a leveraged agreement to take extended-release naltrexone, even though it precludes trying methadone and some other medications for at least a month, is not coerced.

How do we determine which contractual decisions are voluntary and which are the product of duress or coercion? The standard view is that threats coerce but offers do not. And the crux of the distinction between a threat and an offer is that succumbing to a threat would make someone worse off than he previously was ("your money or your life"), while rejecting an offer ("I'll leave you stranded on the highway with your broken-down car unless you pay me $100") will leave the person no worse off than the baseline position he was already in.

Is treatment with naltrexone under these circumstances coerced? Assuming that a person has been fairly charged by the prosecutor, a tendered plea agreement is an offer (which expands the defendant's choices, though they remain constrained), not a threat. In the context of a prosecutorial offer of probation conditioned on taking extended-release naltrexone (in lieu of recommending a more severe sentence authorized by law for the defendant's offense), the defendant who accepts the offer has made a voluntary choice. To be sure, the baseline against which this offer is made (incarceration) is unappealing, and the offender's choice (to plead guilty and comply with treatment each month) has been leveraged by the possibility of imprisonment, but both of these choices are voluntary in a legal sense, as the Supreme Court has properly ruled, and in a moral sense. This is not to say that people so treated won't feel that they "had no choice" and have been pressured to choose an option they might otherwise never have chosen. However, the psychological experience of feeling coerced by circumstances is a phenomenon of everyday life, particularly among people who have few choices to begin with, and it is not the same as coercion in a legal or moral sense.

In a related sphere, addicted physicians who face losing their licenses to practice medicine frequently embrace the option to participate in drug treatment, including use of naltrexone. Interestingly, there is little opposition to offering them these types of leveraged choices. These doctors' circumstances are legally and ethically equivalent to those of addicted offenders, except that one offer is made by a criminal court while the other is tendered, in effect, by a licensing agency. The physicians are faced with the prospect of losing their licenses due to misconduct (their "baseline"), and they are offered the opportunity to keep their licenses by agreeing to treatment with naltrexone; this is equivalent to the offer made to offenders facing a jail term to avoid prison by participating in drug treatment and accepting naltrexone.

Although we believe that a leveraged agreement to take naltrexone is legally and ethically permissible, we also believe that it would be ethically preferable to offer offenders a range of clinically effective drug treatment modalities, including naltrexone as well as other pharmacological (for example, methadone maintenance treatment) and non-pharmacological treatments (for example, drug counseling, case management, etc.). This approach would be properly respectful of the principle of informed choice in medical decisions, which should be honored when possible, even for addicted offenders. (We do not enter into the debate about whether drug addicts make informed, ethically meaningful decisions about their addiction and drug treatment options. It is enough for present purposes to acknowledge that these individuals can make informed, ethically meaningful decisions even if they do not always exercise this ability.)

If naltrexone use were shown to be substantially more cost-effective among criminal offenders than the other treatment modalities, the criminal justice system might be justified in offering only naltrexone as an element of the leveraged agreement. Right now, however, the evidence regarding effectiveness for drug offenders under criminal justice supervision is scanty, not only for naltrexone but also for other pharmacotherapies. That is why it is imperative for criminal justice agencies to commit themselves not only to increasing the availability and use of pharmacotherapies for addicted offenders but also to study the effectiveness of those therapies when they are used to treat this important population.

The No-Choice Approach

The no-choice approach simply orders the offender to participate in treatment whether or not he or she wants to do so; in fact, it might even authorize forcible administration of the medication. For example, a court might sentence someone to a term of probation and order him or her to take naltrexone as a condition of probation, or the parole agency might release the individual and order him or her to take naltrexone, subject to immediate revocation of parole for failure to do so. The development of injectable naltrexone makes it feasible to enforce such an order by administering naltrexone over objection—from a practical standpoint, injections are more feasible to administer over objection than a pill form of any medication; with oral medication, even under mandated conditions, there is some small semblance of volition in swallowing the pill.

Mandated naltrexone injection is ethically and legally problematic, of course, only if the person does not want it and objects to it. If he or she wanted the medication, mandated treatment would be unnecessary but not technically coercive. However, within the framework of medical treatment, and indeed of human rights, any sort of mandated pharmacological treatment over the objection of the individual is ethically and legally controversial.

Compulsory treatment is prima facie unacceptable and must meet a heavy burden of justification. Nevertheless, it is widely accepted that mandated drug treatment can be legally and ethically justified under some circumstances if due process is respected, the treatment is known to be safe and effective, and less restrictive interventions have proven unsuccessful. Examples include antipsychotic drugs for psychiatric patients who lack decisional capacity or who are dangerous to themselves or others, and mandated directly observed treatment for patients with multidrug-resistant TB who are non-adherent under less coercive circumstances. In recent years, several federal courts have upheld forcible administration of antipsychotic drugs as a condition of release on parole.

Mandated naltrexone may also be ethically and legally justifiable under specific circumstances; however, given the current state of the science, mandated treatment with extended-release naltrexone is unsupportable from an ethical standpoint, mainly because there is insufficient evidence of effectiveness. For example, at this point in time it is unknown whether extended-release naltrexone is effective for opioid addiction relapse prevention. To be sure, the formulation would successfully perform its pharmacological action to block the opioid receptor, but how that translates into preventing relapse and re-incarceration is still unknown. Further, in the absence of definitive evidence of effectiveness, residual uncertainty about long-term risks counsels caution, especially as we incorporate lessons learned from delayed understanding of harms from the long-term use of medications such as Vioxx and hormone replacement therapy. And even if the safety and effectiveness of extended-release naltrexone were supported by well-conducted research studies, coercing someone to use it should be regarded as a last resort after other interventions have failed to prevent relapse and reoffending. Even then, coercion may not be justifiable, but before then it clearly is not. And for now we have at least two other policy options that should be tried and evaluated first.

Conclusion

Neuroscientific advances leading to improved understanding of the biological and behavioral correlates of addiction and associated pharmacological developments may well revolutionize the way our society approaches drug addiction. In particular, the development of extended-release naltrexone may prove to be a safe, effective and politically acceptable pharmacological addition to drug addiction treatment options in the criminal justice system. It is possible that the criminal justice system will move rapidly to incorporate use of naltrexone when it is FDA-approved for prevention of relapse into opioid addiction. We hope that criminal justice agencies will subsidize treatment with naltrexone by addicted offenders on probation and parole and will make affirmative efforts to offer naltrexone and other medications such as methadone or buprenorphine to them, either without any connection to their legal status or as a condition of probation or parole in accord with an agreement voluntarily accepted by the offender. Furthermore, as criminal justice policymakers move forward to take full advantage of these advances, it is important to expand treatment opportunities for all people who want to escape the grip of addiction. People with addictions should be able to access treatment when they need it. It would be perverse indeed if committing a crime became the primary gateway to addiction treatment.

CHAPTER 10

Working Later in Life May Facilitate Neural Health

By Denise C. Park, Ph.D.

Denise C. Park, Ph.D., is the director of the Productive Aging Laboratory and T. Boone Pickens Distinguished Chair in Clinical Brain Science at the University of Texas. Her primary area of study is memory in the aging brain. She uses neuroimaging and behavior studies to map the changes in neural pathways throughout life.

Could working past age 65 prove beneficial to neural and cognitive health? Denise C. Park, director of the Productive Aging Laboratory at the University of Texas, suggests that continuing to engage in intellectual activities and new experiences keeps the brain running efficiently. Her theory of "scaffolding" holds that in such situations the aging brain develops new circuits that help people respond to cognitive challenges.

WEALTHIER COUNTRIES are experiencing a rapid and significant increase in their aging populations, thanks to (1) increased longevity; (2) decreased birth rates and (3) the progression of the baby boomer population to older adulthood. Despite tremendous concern in American and other Western societies about the cost of health care and retirement benefits for this aging population, we too seldom ask what the consequences would be for both individuals and society if people continued to work well past 65 and retired at later ages. How would this affect individuals? Would the sustained social engagement and mental activity associated with many jobs and professions result in better cognitive health? What about the effects upon society? Would a longer work life enhance social connectivity across generations, increase mental and physical health in older citizens and, accordingly, lower health-care costs? We must learn more, but I believe that the answer to all of these questions is yes.

We know surprisingly little about the effects of continued engagement in work on neural and cognitive health. Cognitive frailty is almost certainly the most important age-associated health problem that we confront as a society.

We must learn how to slow cognitive aging. Work in my laboratory as well as many others has demonstrated that even healthy older adults show some decline in memory and in the speed at which they process information. Slowing down the process of cognitive aging would help individuals maintain the skills they need to live independently for their entire lives.

Cognitive Engagement May Maintain the Brain

Broad research into mental activity and engagement suggests that continuing employment might be a plus for individuals. Maintaining an active mind or staying deeply engaged in meaningful activities in late adulthood could help sustain a high level of cognitive function. Nearly all the available evidence suggests that an active, engaged lifestyle—both intellectual and social—across one's life span is associated with enhanced cognitive health and a later age of onset of dementing illnesses; large-scale demographic studies suggest that engaging in reading and other intellectual activities is associated with delayed onset of Alzheimer's disease.[1] These findings have two possible explanations. One is that cognitively healthy individuals stay deeply engaged throughout their life span, and the association of engagement with good cognitive health occurs because those who are less healthy slowly disengage over many years. The other, more attractive, possibility is that staying active and engaged actually sustains cognitive health, maintains intellectual vitality and staves off Alzheimer's disease and other dementing illnesses.

Teasing apart the relationship between lifetime mental engagement and cognitive health is not an easy task; scientists can't conduct an experiment in which they randomly assign one group to retire at age 60 and another to retire at age 75. If sustained engagement in work promotes cognitive health, the later retirees should do better than the earlier retirees. Studies do hint at answers.

Evidence Emerges

Carmi Schooler at the National Institutes of Health, using a technique that allowed him to assess causal relationships, found that adults who performed intellectually challenging jobs across their life span showed more cognitive flexibility in late adulthood than those who performed less demanding jobs.[2]

More recent work in the neurosciences also suggests that novel experiences may indeed increase cognitive health and even the size of parts of the brain, providing a neural reserve that could help us maintain function into old age. Animal studies clearly demonstrate that rats or dogs that experience an enriched and stimulating environment, even one introduced in old age, learn better and have greater brain volume than animals in less enriched environments. Research in humans also suggests that stimulation and sustained learning may play a role in brain growth, even with age. One seminal study revealed that London taxicab drivers who had spent many years finding their way through the complex streets of London had a larger hippocampus—a brain area active during way finding and navigation—than age-matched control subjects. Of particular relevance to the question of the impact of work on cognitive aging, the effect of cab driving on hippocampal size was more pronounced with more years of experience, suggesting that sustained work as the cab driver got older magnified the effect.

Perhaps the most compelling evidence regarding the impact of novel experiences on brain volume and function comes from a study at the Max Planck Institute in Germany. Adults with a mean age of 59 spent three months learning to juggle three balls. Although only about half the participants were able to achieve competence in this complex skill, those who succeeded had increased volume in a mediotemporal area of the visual cortex as well as the nucleus accumbens and the hippocampus, suggesting that sustained novel experience can increase the sizes of neural structures. Notably, the changes in the nucleus accumbens and hippocampus were transient, disappearing three months after the juggling ceased. This intriguing study provides clear evidence that continued skill performance is necessary to maintain some gains from experience, and it strongly supports the "use it or lose it" adage.

Drawing meaning from these findings in the realm of work and everyday life in older adulthood is not an entirely straightforward exercise, however. Although structural increases in brain tissue based on novel experience are exciting, they do not mean that continued engagement in the workforce will maintain cognitive health into old age. Few jobs in late adulthood entail continuous challenge and the learning of complex new skills and behaviors, which the studies I've described suggest is important for neural health and growth. In fact, I have argued that most jobs that older adults hold are characterized by "maintenance functions," tasks that rely on knowledge and well-practiced skills rather than active new learning.

Neural 'Scaffolding'

The aging brain may be ripe for the latter—active learning is likely to have a significant cognitive payoff. Patricia Reuter-Lorenz and I have argued that the brain is remarkably adaptive in response to the neural challenges that are part of normal aging. Our Scaffolding Theory of Aging and Cognition (STAC) suggests that the aging brain, when confronted with the joint challenge of declining neural resources and a cognitively demanding task, develops "scaffolds"—new circuitry that helps maintain task performance.[3] Evidence for neural scaffolding emerges from functional imaging studies showing that older adults typically engage more brain tissue than young adults when performing a demanding cognitive task. Most often, this additional activity is in a region in the opposite hemisphere from an area active in young adults, as well ("bilateral recruitment"), or in an area larger than that seen in the young adults ("penumbral activation"). Evidence from my lab and others has provided clear documentation that this additional scaffolding is compensating for areas of the brain that are functioning somewhat less efficiently than in younger adults.

According to the STAC model, neural scaffolding that occurs with age is similar to the development of new circuitry when, at younger ages, we acquire a new skill or learn new information. As we age, scaffolding develops not just in the context of new learning but also to maintain a response when other neural structures or circuitry is no longer sufficiently healthy to meet the challenge of a cognitive task. We suggest that the defining feature of a healthy brain is the ability to continue the scaffolding process in response to challenge. A particularly important tenet of the STAC model is that cognitive activity promotes this ability. On the other hand, the collapse of the scaffolding process is characteristic of various forms of dementia (though the cause of the collapse will be different for different dementia types). With the STAC model as a guide, a critically important goal of late adulthood should be to maintain the ability of the brain to change and respond to cognitive challenge. To the extent that social structures can foster the maintenance of this ability with age, older people might have a higher quality of life and be more able to contribute to the vitality of the economy. Policies that encourage active participation in the workforce into late adulthood, such as providing corporate incentives for new training and incentives to individuals for training and undertaking new challenges, would help.

Following the STAC model, let us explore the cognitive demands of a relatively routine job and the cognitive challenges it might represent. Consider a receptionist for a large medical practice. In order to be effective, a receptionist has to stay on top of scheduling, remember who's who among patients and respond to different physicians' individual preferences. Invariably, new software for patient management, new phone procedures and new insurance forms will be introduced. These considerable memory demands will require scaffolding, and we can see how functioning in such a context might better stimulate one's neural health than would a quiet life at home or even a retreat to the golf course, where events are highly routine and usually require little adaptation to circumstances.

At the same time, neither work nor leisure inherently requires ongoing cognitive challenges, and the demands of different types of jobs can be quite deceptive. One can imagine that something as mundane as being a Wal-Mart greeter could have a significant memory load (e.g., learning the names of repeat customers, keeping special sales in mind each day, etc.) or offer essentially no stimulation or opportunities for scaffolding (e.g., automatically welcoming each customer in a repetitive fashion). Similarly, a leisure-filled lifestyle is not inherently unhealthy—variety, challenge and stimulation are possible. Someone who volunteers as a museum guide and has several active hobbies may experience considerably more cognitive challenge than he or she did in the workforce. The STAC model posits that scaffold development occurs as a result not of work or leisure per se but of situations that are optimally challenging to the cognitive system and that engage many different domains of cognition, including attention, working memory, long-term memory and activation of knowledge systems.

Testing the Scaffolding Theory

Although many aspects of the STAC model are broadly accepted, such as the notion that the aging brain expands areas of activation in response to cognitive challenge, other important facets are more speculative. The model suggests that continuous cognitive challenge and breadth of stimulation in a novel environment would promote more scaffolding and greater brain health. To test this hypothesis, I, along with my colleague Jennifer Lodi-Smith, have developed a learning environment that we call Synapse. We are enrolling research participants in the Synapse environment and providing them with a demanding program that will require 20 hours of their time per week for 14 weeks. Participants will be novices at quilting and digital photography and will be enrolled in a class that specializes in one, the other, or both. They will receive instruction and work on increasingly challenging projects that will stimulate cognitive, motor and social function. We will compare their cognitive performance, brain structure and neural function (as gauged by functional magnetic resonance imaging) to results in individuals who participate in exclusively social activities, in self-paced learning tasks at home or in a control group that does not participate in activities. We hypothesize that the group that learns multiple skills with the combination training will develop the greatest number of scaffolds and will show enhanced attention and memory, as well as more selective recruitment of neural resources. We hope that the skills acquired by Synapse participants will provide continued benefits even after the participants leave the program and that these gains (or at least protection against cognitive loss) will endure.

If older people have jobs that include cognitive challenges similar to those we are creating, our research suggests that they may be able to maintain a healthier brain. Although we have much to learn about the aging mind and how to preserve its vitality, new imaging tools are allowing us to take giant steps as we examine these questions further. One of the premier challenges of the 21st century lies in determining what behaviors will protect neural health and then developing public health initiatives to encourage these behaviors in our communities. Sound social policies that encourage older people to keep working will have direct benefits to our economic system. Such policies also could be neuroprotective, resulting in later onset of dementing illnesses, an outcome that offers gains for society thanks to reduced caregiving and health-care costs, as well as extended time with beloved family members.

CHAPTER 11

Managing Conflicting Interests in Medical Journal Publishing

By Adam F. Stewart, S. Claiborne Johnston, M.D., Ph.D., and Stephen L. Hauser, M.D.

Adam F. Stewart is a medical journal publishing professional. He has published on a broad range of topics, including the neurological impact of combat injuries, and has held editorial positions with the American Journal of Industrial Medicine and Pediatric Pulmonology. Currently he is the managing editor of Annals of Neurology.

S. Claiborne Johnston, M.D., Ph.D., is a professor of neurology and epidemiology and the director of the Stroke Service at the University of California, San Francisco. His research has focused on the prevention and treatment of stroke and TIA. He is also the executive vice editor of Annals of Neurology.

Stephen L. Hauser, M.D., is the Robert A. Fishman Distinguished Professor and chair of the Department of Neurology at the University of California, San Francisco. His research has advanced our understanding of the genetic basis, immune mechanisms and treatment of multiple sclerosis. He is also the editor in chief of Annals of Neurology.

Editors of scientific journals are the gatekeepers for much of what we know about science and medicine; what appears in their pages often turns up later in drug advertising, advocacy group fundraising, patients' Internet search results and science news in the daily media. As editors, they must ensure the accuracy and novelty of the information they disseminate. Cerebrum invited the editors of a top scientific journal, Annals of Neurology, to reveal their perspective on the difficulties of dealing with the conflicting interests of authors, the pharmaceutical industry and journals themselves, all while maintaining the quality of the information that reaches the public.

Where do we obtain our facts as well as our theories? Both are being published daily in the medical journals we read.... Who decides what we read? The editors.[1]

—JAN P. VANDENBROUCKE,
FORMER EDITOR IN CHIEF, THE LANCET

EVER SINCE THE JOURNAL DES SAVANTS, the first academic journal, began disseminating the Philosophical Transactions of the Royal Society of London in 1665, scholarly journals have been an essential engine driving scientific inquiry, investigators' careers, industry profits and the agendas of nonprofit organizations. Most scientists work in highly competitive, hierarchical environments in which creative output is the currency of trade and an individual's reputation and rank are determined by productivity as measured by the number of manuscripts one publishes. The placement of work in top-ranking journals has become increasingly essential to enhance the visibility of scientists' discoveries, promote acceptance by colleagues, attract trainees to their laboratories, secure grant funding and ensure job security. As editors of a clinical neurosciences journal, Annals of Neurology, we find that the urgency for investigators to publish in top journals places great pressure on journal editors, who not only serve as arbiters of quality and taste but must also successfully align the different interests of investigators, peer reviewers and the journal itself with the interests of the readership at large.

Determining Novelty

One challenge we face as editors is to ensure that the submitted articles are reporting on new results. This involves policing duplicate publications and determining exactly what constitutes duplication. It is by no means an easy task; the National Library of Medicine's medical literature and retrieval system (MEDLINE) database currently indexes more than 18 million records published in 5,246 medical journals. In 1969, Franz Joseph Ingelfinger (1910–1980), editor in chief of the New England Journal of Medicine, established what became known as the "Ingelfinger rule," essentially stating that the journal would not consider a manuscript for publication if it had previously been published elsewhere.[2] Although some have claimed that the rule has had a stifling effect on the dissemination of scientific research,[3, 4] it is now standard policy for nearly every medical journal, across all specialties. The distinction between the justifiable practice of splitting data from a major study for publication in multiple journals catering to different specialties and a clear violation of the Ingelfinger rule is often ambiguous and difficult to determine, as it depends largely on the degree of data overlap.

As one Annals of Neurology reviewer mentioned to us recently, some of our more prolific colleagues do seem to be disseminating their work in "the smallest publishable units." Such instances of apparent duplication are relatively benign compared to republication. In fact, considering that each and every citation may add value in today's difficult funding climate, it is perhaps surprising that splitting data is not a more common practice. Other examples of potentially "benign" duplication might be in the areas of review articles based upon a scientist's body of work, interim clinical trial reports or abstracts. Sadly, we believe that some contributors have exceeded any reasonable threshold of benign duplication, crossing the boundary of auto-plagiarism, copying entire paragraphs from prior reports. In our experience, most violations involve the submission of similar review articles to different journals. The proliferation of electronic tools at our disposal to discover and organize content in the medical literature notwithstanding, it remains a challenge for journal editors to discover when the Ingelfinger rule has been broken.

The PubMed search engine remains one of the best tools for the task of detecting potential duplicate publications. In addition to peer review, which we believe is a very good means of identifying novelty and importance, we have adopted a policy at the Annals that no manuscript is accepted without first confirming the study's novelty by conducting a MEDLINE search to look for publications or pre-existing knowledge that belies the novelty of the paper under review. We also use the Google Scholar search engine extensively to screen for prior use of key phrases or paragraphs in a submission under review. On more than one occasion, we have faced the unpleasant task of notifying an author that our search revealed that identical content had already been published elsewhere.

The Computational Biology Group at the University of Texas Southwestern Medical Center at Dallas has developed another tool, called eTBLAST, specifically designed to identify similar articles in the biomedical literature by combing through abstracts of published articles. A user enters a string of text, and the program searches for similarities in the published literature. A recent study published in Nature using the eTBLAST technology estimated that, of the approximately 18 million citations in the MEDLINE database, nearly 200,000 are duplications—most of them instances of auto-plagiarism.[5]

Unfortunately, neither PubMed nor eTBLAST is capable of detecting auto-plagiarized work that has not yet been published. Remarkably, even in 2008, journal editors remain critically dependent on the honesty and integrity of authors and peer reviewers to protect against plagiarism and other forms of misconduct.

Many investigators also feel pressure to present findings to scientific audiences before their work is published in a peer-reviewed journal, both to scoop competitors and to obtain greater acknowledgment of their work. Drug or device manufacturers are also interested in disseminating positive trial results as quickly as possible, and publicly traded companies have a legal responsibility to immediately notify investors of any news that might materially affect the value of their holdings. Authors often ask us whether presenting data at a scientific meeting constitutes duplication in the literature and precludes publication in the Annals. If not, to what degree does public discussion affect a paper's novelty? Such questions are never easy to answer.

The question of whether it is acceptable to discuss data in a public forum before results are published in a peer-reviewed journal has generated a good deal of confusion. For the record, our position is that scientists are allowed to discuss their data before it is published, and in many instances they have a moral responsibility to do so. Rapid presentation of clinical trial results is clearly in the best interests of patients. From an editor's perspective, the impact of a paper may be diminished if the findings are already widely known; however, a paper is considered novel as long as the findings have not been previously published in the primary peer-reviewed literature.

Limitations of Peer Review

Peer review is meant to protect readers and editors from the biases of the investigators submitting their manuscripts, but peer review also introduces problems of its own, including potential conflicts of interest. One of the more common complaints that we hear around the Annals editorial office is that peer review is too slow. Authors often grumble about the possibility of data theft from competitors who may be serving as opportunistic peer reviewers in order to purposely delay the review process while they rapidly prepare and submit similar data of their own to another journal. One author explicitly requested that we not send a submitted paper to "any expert in the field," for fear that the referee might steal the data. Although the vast majority of such complaints can be explained by anxiety and eagerness to publish, it is not entirely uncommon for a peer reviewer to take weeks and sometimes months to complete the assigned review. In such instances, we sometimes wonder if some of our most concerned authors might be legitimately suspicious after all. No system can be perfectly fair, and the process of peer review is no exception.

The biases of peer reviewers are largely unpredictable and can occur in either direction. A reviewer can stymie a competitor by delaying review, embrace a validation of his or her own work by providing a laudatory evaluation with scant criticism or suppress dissent by harshly critiquing a paper that presents evidence contradictory to that of the reviewer's work—no matter how convincing or rigorous the data.

Last year, for example, the Annals received a paper that described a successful animal study of RNA interference as a potential therapy for a neurodegenerative disease. The paper was sent out for peer review, and shortly thereafter one of the reviewers contacted the editors to point out that the reviewer had presented data at a scientific conference that were suspiciously similar to those in the paper, and that these data (the reviewer's) were currently in press with another journal. Her data and those in the submitted paper differed in only one detail. After pointing out this highly remarkable coincidence, the reviewer spent 35 days reviewing the paper—she protracted her review intentionally to ensure that her own paper went to press first. To her credit, she told us openly that she was doing so.

Regrettably, there is little that editors can do in such instances without compromising the anonymity of the reviewer or accusing the author of appropriating an idea without much credible evidence to support the accusation. When the data are solid and the paper is favorably reviewed, the editors are obligated to publish.

In another example, the Annals published a paper describing a new animal model for a different neurodegenerative disease. The reviewers found that the study was well executed but, within weeks of the article's publication, we received information from other scientists that the work possibly was flawed. Several months later, we published a series of papers from authors at various institutions around the globe, most reporting a failure to replicate the original work. Although this process was painful, it highlighted for us a responsibility that we believe all journal editors must assume: as a general rule, we will publish any credible report that calls into question content previously published in our journal.

A final example: we recently published the discovery of a possible virus linked to multiple sclerosis. In doing so, we were aware of similar research in the past that had failed to identify this common virus (varicella, known to cause shingles and chicken pox). In this new report, however, modern molecular methods were used to search for the virus, and the data as presented were so clear-cut and definitive that the findings could not be dismissed. Because we were highly skeptical, given the many earlier reports of virus isolations in multiple sclerosis that were not reproducible, we insisted on a blind replication using coded samples sent from another source. The authors performed this replication and confirmed their initial finding. Although we continued to suspect that the findings were unlikely to be true, we decided to publish, with an accompanying cautious and somewhat skeptical editorial. Not surprisingly, we are currently reviewing a paper from another lab that failed to replicate.

Carl Sagan's famous axiom that extraordinary claims require extraordinary proof would seem to be a useful rule for all journal editors. In a larger sense, J. P. Ioannidis and others have raised the concern that many claims in the biomedical literature are not reproducible.[6] Human studies that involve large-scale genetic, proteomic, or immunologic investigations have the particular risk of type 1 errors (statistical errors better known as "false positives"). Because of that, we have adopted a firm editorial policy requiring confirmation of such studies with an independent data set.

Despite our best efforts to guarantee scientific rigor and reproducibility, to temper overinterpretation of data, to tone down overstated conclusions, and to sober overzealous authors, certain controversies are simply unavoidable, and the possibility of bias will always exist. The pressure to publish can be great indeed, and, when the journal's considered disposition on a paper is unfavorable, authors work hard to convince us otherwise. In the final analysis, we are responsible for setting the bar at an appropriate level.

Embargo and the Lay Media

Embargoes are another frequent source of ambiguity and conflict. Many of our authors' institutions and funding bodies are eager to disseminate press releases when an article is accepted for publication; however, the policy of most journals is that the press release is subject to an embargo so that the primary peer-reviewed literature is able to publish first. Part of the reason for embargo is to protect the journal's revenue; once the essential findings are available in the public domain, the original literature has less monetary value. More important, however, is that the primary literature should publish first so that data are available to the public before the lay media are allowed to discuss results from a scientific study.

Once an embargo is lifted, media coverage often falls prey to inaccuracy and errors of omission. A 1991 New England Journal of Medicine study of lay media coverage of three medications used to prevent major diseases revealed that the media had inconsistently and inaccurately reported the benefits and hazards associated with the medications.[7] In 2003, the Canadian Medical Association Journal published a similar analysis of 193 newspapers covering five new drugs, revealing that 100 percent of the articles reported at least one therapeutic benefit, but 68 percent failed to report any harm.[8] Both studies found that financial conflicts of interest between the pharmaceutical industry and key expert sources cited in the reports were almost never disclosed.

Financial Conflicts of Interest

Disclosure of financial conflicts of interest is an absolute requirement in medical journals in order to diminish the potential role of industry incentives. Everyone in the academic community is aware of numerous instances of high-profile investigators who failed to disclose significant conflicts of interest, leading to personal and professional embarrassment and, for some, representing serious violations of the law. Most of these lapses in judgment concern failures to fully disclose financial relationships with pharmaceutical or device manufacturers, or with financial firms.

Ghostwriting, writing by someone who is not listed as an author, is one common example of an unhealthy relationship with a commercial sponsor. At the Annals, we continue to receive manuscripts that appear to have been ghostwritten as part of a marketing agenda, perhaps by a drug-company employee, often with the purpose of encouraging physicians to prescribe drugs or devices for uses that are not included in the product's labeling. This "off-label" prescribing is legal, but pharmaceutical companies are not legally allowed to promote off-label use through advertising.

Whenever we suspect that a manuscript reflects in any way the hidden agenda of a commercial entity, we reject it—even when the authors (usually distinguished senior investigators at more than one institution) lobby for a reversal of the verdict. Finally, it is important to recognize that relationships between industry and investigators are not always financial ones, and that non-financial entanglements—perhaps involving access to an emerging technology or leadership in a clinical trial—may create very powerful relationships that can lead to bias and often are not formally disclosed.[9]

As editors, our most inviolable charge is to ensure that the content of our journal is of the highest possible quality, that we always operate in the public's best interest, and that we never give readers cause to question our independence. Conflicts of interest can appear in many forms, both obvious and subtle. Only by recognizing that conflicts of interest will always exist, that human beings are fallible, and that scientific inquiry is never free from potential bias can we make certain that such conflicts are recognized and managed in a consistent manner and under the full light of disclosure.

A generation ago, the science historian Derek de Solla Price wrote that 80 to 90 percent of all scientists who have ever lived are currently alive,[10] a situation that likely remains true today. Modern scientific journals occupy a position of sacred trust in this crowded world as they endeavor to assess new discoveries and disseminate novel findings to the community at large. As are scientists, journals are extremely competitive with each other as each vies for supremacy in its respective area of inquiry. For journal editors, juggling the competing demands, deadlines and pressures of modern scientific publishing while remaining free from bias, fair to authors, attentive to quality and true to the highest ethical standards is a daunting day-to-day challenge.

CHAPTER 12

Pediatric Screening
and the Public Good

By Jennifer Kwon, M.D., M.P.H.,
and Richard H. Dees, Ph.D.

Jennifer Kwon, M.D., M.P.H., is an assistant professor
in neurology and pediatrics at the University of Rochester.
Her research focuses on neurogenetic disorders, including
neurodegenerative disorders and neuromuscular disease.
She also evaluates infants and preschool children with
developmental delays, including speech disorders and autism.

Richard H. Dees, Ph.D., is an associate professor of
philosophy, neurology and medical humanities at the University
of Rochester. He is interested in the ethical implications of
using medicine to make us better than we are, particularly
in terms of brain functions such as memory, cognition
and personality.

The advantages of screening for diseases and disorders in children seem obvious. Ideally, tests catch problems early and increase opportunities for treatment and recovery. However, Jennifer Kwon and Richard H. Dees note that screening programs can have a number of complications, including ambiguous benefits, the need to educate families and the public, results that land in a gray area between normal and certain disorder, blurred lines between screening and research and competition for scarce funding. Kwon and Dees urge caution and careful consideration of potential costs alongside potential advantages.

AT FIRST GLANCE, the idea of screening infants and young children for diseases is a sure winner. Early detection of diseases can help families begin treatments that can prevent irreversible disability or death, so mass screening programs are a public health boon. By allowing society to treat illnesses before they manifest themselves, we keep patients from much harm and pain and save money by creating a healthier population and reducing long-term care needs. However, to evaluate these screening programs, we must understand the specifics of their costs and benefits, in terms of both the monetary implications and the effect on the quality of people's lives. Just as important, we must look at the potential effects on everyone who might be affected by these programs.

The classic screening success story is the development of the nationwide program of newborn screening for phenylketonuria (PKU). PKU is an inherited disorder affecting 1 of every 15,000 infants in the United States. In PKU, the essential amino acid phenylalanine accumulates in the brain, leading to early brain injury and eventually to severe seizures and mental retardation. Dietary restriction of phenylalanine can prevent neurologic damage, but only if instituted in the first months of life. Therefore, early detection allows children with PKU to live productive lives. The test is easy to administer, accurate, and low-cost even when administered to all newborns; treatment is readily available, and the payoff is extraordinarily high, since it prevents severe mental retardation. As a result, the PKU testing program is a paradigm for why and how we screen for diseases.

On the flip side, the classic anti-screening story is the disastrous attempt to screen for carriers of sickle-cell anemia among African American children in the 1970s. Carrier screening identifies individuals who are unaffected themselves but who may nevertheless pass a genetic disease on to their offspring. At the time, there was no cure for sickle-cell disease, so carrier screening might have been a useful means to help people make reproductive decisions to reduce the incidence of the disease.[1] Indeed, a carrier screening program that began about the same time among young adults in the Ashkenazi Jewish population reduced the incidence of Tay-Sachs Disease, an always fatal neurodegenerative disease found almost exclusively in this population, by 90 percent. However, the sickle-cell test was aimed at African American children, rather than at the adults who could have used carrier information to make reproductive decisions, and the test was understood so poorly that even the U.S. Air Force refused to allow mere carriers to become pilots because it was assumed that they had the disease and were in danger of passing out in the cockpit. The program thus benefited no one, and it was charged that it led directly to racial discrimination.

Despite such cautionary tales, screening programs for infants and children have proliferated as new technologies have made it possible to test for many rare diseases from a single blood sample. For that reason, the American College of Medical Genetics (ACMG) now recommends mandatory newborn screening for 29 diseases.[2] The vast majority of these disorders can cause irreversible neurologic injury over time if not identified and treated promptly. Some disorders, such as PKU and hypothyroidism, are effectively treated by dietary management or medications or both. In other cases, such as cystic fibrosis, the disorders are not presently curable, but early and meticulous medical management can lead to longer and more productive lives. In addition to screening newborns, new programs have been developed for other diseases that emerge during childhood. Recently published guidelines from the American Academy of Pediatrics, for example, suggest that children be screened for autism at regular intervals, beginning at 9 months, so that early treatment can begin and thus "minimize the core features and associated deficits, maximize functional independence and quality of life and alleviate family distress."[3]

Advocacy groups call for more widespread pediatric screening, recommending testing for more than 50 disorders, most as part of the groups of tests done on newborns. This higher number is in part a reflection of improvements in technologies for diagnosing and treating rare disorders since the college's 2005 recommendations, but it primarily reflects the different criteria that the advocacy groups use to judge the effectiveness and the value of the tests for particular diseases. These criteria vary among the different constituencies affected by screening: public health officials, families who have experienced the pain of lengthy diagnosis and failed treatments in rare disorders, and even biotechnology companies that stand to gain by prompt treatment of certain rare disorders.

Sensitivity versus Specificity

The accuracy of the tests and of the laboratories that analyze the results affects the value and usefulness of screening. In general, screening programs are designed to identify the people who have a disease (true positives) and to exclude those who do not (true negatives). But no matter how good the screening test, there will always be those incorrectly identified as having the disease (false positives) and those incorrectly identified as not having it (false negatives). A perfect test would identify all children who have the disease and only those children, but, in the real world, a tension often exists between identifying everyone who has the disease (the test's sensitivity) and excluding all those who do not have it (the test's specificity). In other words, to minimize the number of false negatives, we often accept more false positives. To assess the costs and benefits of any screening program, then, we must consider these four categories:

1. *True negatives.* The vast number of children screened would fall into this category, since most diseases do not have a high prevalence in children. After the test, the parents of these children know that their child does not have the disease for which he or she has been screened, and so they feel confident that they do not need to worry about it. With screening they gain some peace of mind—a small

benefit, albeit one that affects a large number of people.

2. *False negatives.* For most screened diseases, the costs of a false negative are grave: where screening is justified, early treatment is essential to avoid severe disability or death, and the programs are explicitly designed to have no false negatives. If there are any such "missed" diagnoses, the consequences could be devastating—even if the infant being screened is in the same condition he or she would have been in without the program. Parents would not know the infant needed treatment. Worse yet, the test result may give families and even physicians a false sense of security that makes them slower to diagnose an affected infant.

3. *False positives.* Most of the otherwise avoidable adverse impacts of screenings fall on people who do not have a disease but who are identified by a test as having it. There is some risk for normal individuals to be subjected to treatments that are unnecessary or harmful. Therefore, it is important, though time- and resource-consuming, to perform confirmatory tests and counsel worried parents about the disease. These costs are not trivial. In any rare condition, finding the true cases of disease will result in a large number of false positives. So even for PKU, the number of false positives swamps the number of true positives. In New York, for example, 273 of the 240,000 newborns tested in 2007 had a positive result for PKU, but only 13 actually had the disease.

4. *True positives.* These patients obviously benefit the most from testing. In 2007 in New York alone, 655 newborns were discovered to have serious diseases for which immediate treatment was provided. Often money is saved as well, since the costs of early treatment are much lower than the costs of care for severely disabled patients.

We must also factor in the effects of screening programs on the public at large. The most obvious effect is monetary, since these programs cost taxpayers money—money that could be used for other purposes.

Given this framework, the basic argument for screening is usually that the benefits of early detection and early treatment for the true positives outweigh the anxiety and the occasional mistaken treatment of false positives and the costs of setting up and maintaining the program itself. In the cases in which early treatment has a clear impact on the children affected, such as PKU, such a calculation easily favors screening.

Potential Pitfalls

Few cases are so simple, though, for one of at least five reasons. First, the benefits of early detection may not be so clear. One of the factors that motivated the AAP recommendations for routine autism screening, for example, was the increasingly convincing evidence that early intervention—before age 4—significantly improves outcomes. However, the recommendations assume that routine screening—costly for such a large group of children—will discover a significant number of children with autism that normal checkups would miss. These otherwise-undiagnosed children will then receive services at a time when they are likely to be most effective. But the costs of treatment may be high, since early interventions usually involve intensive and very expensive "teaching" or "behavioral training." Moreover, for some, these interventions are unsuccessful, and still others will receive this expensive therapy because they screened positive but never would have developed the more severe symptoms. Because increasing the functional abilities of autistic children even by a small amount can significantly improve their lives and those of their families, autism screening seems justified on balance. But coming to that conclusion is much more complicated than it is for PKU.

Second, the need to educate both parents and the public about the testing and the disease complicates the basic argument for screening. Every parent should have some basic understanding of what tests their children undergo, and of course the parents of children who test positive require careful explanation of the results, the disease their child might have and the options for future courses of action. In addition, the general public must be informed through various media about the tests and what they mean. Without such education, the screening can actively cause harm; indeed, if a diagnosis carries a stigma, as sickle-cell anemia did in the 1970s and as autism still does among some, then an extensive public education program is needed to overcome that misinformation and the resulting resistance to testing. Such education, then, is essential for success, but it also increases program costs.

Third, testing often reveals patients who have "abnormal" results but who do not clearly have the condition in question. For example, autism screening is likely to find children who lack important social skills but who do not clearly fall on the autism spectrum. We expect such children to benefit from increased services—but then, almost any child is likely to learn better with increased one-on-one attention and a careful scrutiny of his or her educational needs. Such "marginal" situations might also subject children to whatever stigma might be associated with a diagnosis. To take another example, consider the novel newborn screening program in New York State for Krabbe disease, a highly debilitating neurode-generative disorder that leads to death in early childhood; it is caused by a deficiency in galactocerebrosidase, one of the enzymes needed for normal neuronal development. New York state, via the test, has identified newborns with low enzyme levels but without a severe enzyme deficiency, raising concerns that they may develop the disease later in life. No one knows exactly what these results mean in terms of prognosis, however. By monitoring the infants, we may discover whether they ever develop the disease and may be able to determine what enzyme levels correlate with eventual disease development, but efforts to do so have turned the screening program into a statewide clinical research project.

Indeed, in some cases, advocates for children with rare disorders may promote screening programs to invigorate research as a means of finding a cure. This is a fourth complication in the argument for screening. Krabbe screening was implemented in New York after intense lobbying from former Buffalo Bills quarterback Jim Kelly, whose son Hunter died from the disease. Currently, the only hope for a cure lies in the early use of umbilical cord stem cells. This treatment, however, involves chemotherapy and stem-cell transplantation in infants as young as 1 month old, since the treatment is effective only before neurologic symptoms appear. Because the experimental treatment is new, the long-term prognosis for those receiving the treatment is unclear, and the transplant itself is risky at such a young age. In the first full year of the screening, four children were found to have been at high risk for developing the infantile form of Krabbe disease; of those, two had transplants and one has died from transplant-related complications. Since Krabbe is fatal, we might think such experimental therapies are worth pursuing, and of course parents who disagree are free to refuse the treatment on behalf of their children. But the screening program now serves as a state-sponsored recruiting opportunity for Krabbe disease research.

At a minimum, the situation has strayed far from the ideal of a screening program such as that for PKU. We are now using legal mandates that compel parents to decide whether or not to allow their newly diag-nosed children to participate in experimental treatments whose outcomes are still unknown.[4] Such research programs may be legitimate, but they require a different kind of ethical justification. Essentially, the advocates of widespread screening are leveraging general public support for screening of curable diseases to advance their own research agendas. Their desire to do so is certainly understandable. For diseases such as Krabbe, a screening program may be the only means by which participants for clinical trials can be found. But the state cannot pretend that the testing is needed so that a proven treatment can be provided to those afflicted with disease. Since the program is research, it needs to be treated like research: it needs oversight from a group of individuals who are free of ties to advocacy groups that are promoting further testing or to surgery centers and to drug companies that provide treatment. In the case of Krabbe testing in New York, the state failed to set up any oversight board, and the informal group that has emerged has close ties to Kelly's advocacy group, Hunter's Hope. In sponsoring the oversight group, Hunter's Hope has generously provided what the state should have established and did not, but the help has come at the cost of objectivity.

The basic argument for screening is complicated if we take into account the interest of the public at large. As the use of screening primarily for research purposes shows, all such programs must be evaluated in the context of public health in general. Insofar as any screening program uses limited resources, it must be evaluated not just on its own merits but also relative to other uses to which those resources might be put. Screening mandates rarely come with separate funding; therefore, most of the cost of the screening and follow-ups is drawn from a larger pool of money. Thus, to fund screening, cuts will be required in programs that do not have the same political cachet. In Mississippi, for example, the funding for increased newborn screening came at the same time as cuts in Medicaid, and infant deaths rose by 65 the following year.[5] While a direct link between these two occurrences cannot be established, their correlation highlights the reality that, whenever resources are limited, we will have to make hard decisions about which programs to fund. As a society, we often pretend that we do not have to make such choices, but until we create a different health-care system, increases in screening will affect other programs, and we will have to decide which of many worthwhile programs to eliminate.

Despite these caveats, screening programs can still be a public health boon. The key is to avoid getting caught up in the technological advances that make more tests possible or in the advocacy politics that create demand for more diseases to be tested. Instead, we must weigh carefully what we think a screening program can accomplish and what its costs will be, both for the people tested, whether or not they are affected, and for the public at large.

CHAPTER 13

Discovering That Rational Economic Man Has a Heart

A Cerebrum Classic*

By Lee Alan Dugatkin, Ph.D.

Lee Alan Dugatkin, Ph.D., professor and distinguished university scholar of biology at the University of Louisville, specializes in the evolution of social behavior in animals. He is the author of The Altruism Equation (Princeton University Press, 2006) and Cheating Monkeys and Citizen Bees: The Nature of Cooperation in Animals and Humans (Free Press, 1999).

*From Volume 7, Number 3, Summer 2005

Author's Note

When I wrote Discovering that Rational Economic Man Has a Heart in 2005, I did so because it seemed to me that the new science of neuroeconomics had great potential, both theoretical and practical. That a single field would draw inspiration from neurobiology, evolutionary biology, economics and psychology was, to say the least, exciting. In 2005, neuroeconomics was a toddler and, like a toddler, it soaked up and absorbed information from the world around it, sometimes making silly mistakes along the way. Without stretching the analogy too far, I think it is fair to say that neuroeconomics today is more like a young man or woman about to enter graduate school—mature and poised, brimming with ideas and ready to get very serious about *testing* those ideas. Often some of the very best work a scientist ever does comes in the graduate school years; let's hope that this is equally true for the field of neuroeconomics. Personally, I am cautiously optimistic.

What happens in our brains when we make an economic decision? Do we coolly weigh alternatives to discern and act on our self-interest? Generations of economists and policymakers have thought so. But the traditional model of rational economic man must now be reconciled with the discovery by neuroscientists that when we make an economic decision the areas of our brain associated with emotions are also active. Can rational economic man be saved?

IF WE WERE TO NOMINATE A POSTER BOY for studies on brain and behavior, the winner would be Phineas Gage. In the summer of 1848, Gage was in charge of a crew laying track for the Rutland & Burlington Railroad. One of his tasks was to use gunpowder to blow up rock sections, and, in a tragic error, Gage accidentally tapped on the gunpowder with an iron rod—and made history. The explosion propelled the three-foot iron rod upward through Gage's cheek, then through his brain, and out the top of his head. Gage not only survived but also within minutes was speaking and walking. Within two months, doctors pronounced him healed.

Indeed, Gage seemed physically intact, particularly so for a man who had an iron bar shot through his head. But he was, in one important sense, anything but cured. His personality had changed in a dramatic way, so much so that people who knew him before and after the accident would comment that "Gage was no longer Gage."

Before the accident, Gage was an amicable fellow, but that changed completely. One of his physicians, Dr. John Harlow, described the changes thus: "The equilibrium or balance, so to speak, between his intellectual faculty and animal propensities" was gone. He displayed "little deference for his fellows ... [and was] capricious and vacillating, devising many plans of future operation, which were no sooner arranged than they were abandoned." It was as if Gage had lost his ability to plan and act in a reasonable manner, particularly when it came to social interactions. Could a brain injury that by all accounts should have killed Gage instantly really have resulted in such a specific sort of personality problem? All that is known for certain is that the iron rod in question traversed his prefrontal cortices.

Is that, then, where our abilities to handle social interactions lie? If so, in what parts of the frontal lobe, and what does this tell us in general about how the brain works?

Before attempting to answer these questions, let us look at another case history. In his wonderfully entertaining and thought-provoking book "Descartes' Error," Antonio Damasio, M.D., Ph.D., tells us of a modern-day Gage, named Elliot. In Elliot's case, a large tumor wrought havoc in his right frontal lobe, but the resulting changes in his personality were remarkably similar to those noted in Gage. Elliot's memory was fine, his mathematical abilities were unchanged, his language was intact and his perceptual skills were seemingly unchanged. Yet socially, Elliot was a disaster. A battery of tests found that, although Elliot was aware of social conventions—what was viewed as acceptable and unacceptable—in real-life social situations, he was simply unable to decide among the countless options and so failed miserably. For example, if he was asked to describe what might happen if he was given too much money by a bank teller when cashing a check, he responded normally. Or, if asked how he might calm down a friend or spouse whom he angered, Elliot again did fine. But faced with the same circumstances outside the controlled environment of the lab, when any decision he might make would have true social consequences, Elliot was hopeless. He would simply freeze up, incapable of choosing among the myriad options available.

In his report on Elliot's behavior, Damasio notes that Elliot seemed almost devoid of emotion and spoke with a calculated cold-bloodedness. Damasio and other neuroscientists have noted that when a patient's lack of social skill is juxtaposed with a complete lack of emotion, the trouble usually stems from problems in the prefrontal cortices of the brain, in particular the ventromedial area of the prefrontal cortices.

From this work Damasio has generated a fascinating hypothesis. Although it is often thought that to make rational social decisions emotion needs to be removed from the process, brain science suggests that without emotion people fail utterly in the social realm. Emotion and rational decision making—that is, reason—are not separate phenomena. On the contrary, they appear to be integrally linked in the human brain.

Studies such as those on Gage and Elliot have informed not only neuroscience but also what has been called "the dismal science"—economics. Although some economic decisions are made outside a social context, those are the minority. Social dynamics, many economists believe, are at the core of economic decision making—that is, decision making about resource acquisition and expense allocation. What I decide affects you, what you decide affects me, and, even more to the point, I care how I fare economically compared with how you fare. Thus, over the past 5 to 10 years, economists and brain scientists have come together and created a new subdiscipline—neuroeconomics—that calls both neurobiology and economics home.

"Neuroeconomics." The word itself has an ominous futuristic ring, conjuring up images of Big Brother engaged in economic mind reading. Indeed, an element of mind reading, or at least brain reading, is involved. Neuroeconomists want to understand what happens in the human brain during economic decision making.

Imaging and the Ultimatum Game

How is an experiment in neuroeconomics conducted? Typically, subjects who are connected to a functional magnetic resonance imaging (fMRI) or positron emission tomography (PET) scanning machine make some sort of economic decision while scientists observe and record their brain activity. Some of the clever scenarios used in these studies predate brain scanning. Economists, of course, came up with the rational-man model of decision making long before brain scans, and psychologists have investigated learning and decision making for more than a century. But brain-scanning technology affords scientists an unprecedented glimpse into the brain as it makes decisions; without this technology, neuroeconomics would not exist.

The decisions that subjects must make are often cast in terms of cooperation, cheating or punishment, which means the studies tie together brain neurobiology, economics and social psychology in novel ways. The evolution of cooperation, cheating and punishment is important in evolutionary biology, too, so neuroeconomics pulls yet another discipline into its fold. Although this new interdisciplinary field today focuses on economic decision making, its approach can be applied to studying all kinds of decisions, such as how people informally estimate probabilities and how people categorize human faces as trustworthy or not. This suggests a future for neuroeconomics in shaping a comprehensive theory of decision making.

A few current studies will give us a feel for what neuroeconomics can tell us about the human brain and decision making. Keep in mind, though, that new studies come out weekly and that new centers for neuroeconomics continue to appear at universities. Here, I can only set the stage for discussion of some of the deep questions posed by neuroeconomics.

First, consider what researchers call the Ultimatum Game. Two people are involved in an economic transaction over a resource that is of value to both (for example, a pot of pretend money), but only one person—called the proposer—controls that resource. The structure of the game is such that if the proposer offers some proportion of the resource to the other person—call him the responder—and that offer is accepted, then both people get to keep their portions. If the offer is rejected, then neither person receives anything. For example, imagine there is $10 in Monopoly money in the pot. Suppose that John (the proposer) offers Jim (the responder) $5. If Jim agrees, each player gets $5. But if he refuses, both John and Jim get nothing.

Let us analyze this game from the traditional perspective of the rational economic man (*Homo economicus*), which holds that a person ranks potential outcomes and acts, in the lingo of economists, to "maximize some utility function"—in this case, taking as much of the pot of money as possible. From this economic perspective, the proposer should make the smallest offer that he believes the responder will accept, for that will obviously leave the proposer with the most money at the end of the game. The responder should accept that offer; if he does, he gets whatever the proposer offered. Otherwise, he gets nothing.

Yet the results tell another story. Most often, in these sorts of behavioral economics experiments, the proposer offers half the resource, rather than a small slice of it. If the offer is not half the resource, the responder often rejects it, even though by doing so he ends up with nothing. For example, Alan Sanfey, Ph.D., and his colleagues at Princeton University examined the Ultimatum Game with 19 subjects in the role of responder and used fMRI to observe their brain activity. They found that when unfair offers (defined as those of less than half the resource) were made, responders often rejected them. As they did so, the area of their brains associated with negative emotional states (in this case, the bilateral anterior insula), rather than those associated with complex cognition (in this case, the dorsolateral prefrontal cortex) were most active. The more the offer deviated from fair, the more active was the bilateral anterior insula when such an offer was rejected. Anger at being treated unfairly by other players appeared to override rational economic reasoning. In the minority of cases when the offer was accepted, the dorsolateral prefrontal cortex was most active.

Sanfey and his team took their experiment one step further. They had the same subjects play the Ultimatum Game against a computer that did exactly what the subject's human partner had done. In a testament to the fine-scale social distinctions that humans make, Sanfey and his colleagues found that subjects were more likely to accept an unfair offer from a computer partner than from a human partner, and activation of the bilateral anterior insula was lower when unfair offers were made by the computer. In other words, although the monetary calculations were exactly the same in both conditions—and hence a rational person should respond similarly in both contexts—subjects were much more likely to view an unfair offer from another person as a violation of social norms and hence respond emotionally.

In an interesting twist to the Ultimatum Game, Erte Xiao, Ph.D., and Daniel Hauser, Ph.D., added the possibility for the responder to attach a written note to be read by the proposer at the same time that the proposer learned of the responder's decision. Because the note was written after the proposer made his decision (and hence no negotiations were involved), it functioned only as an emotional vent for responders. Xiao and Hauser hypothesized that, when the responder was offered an unfair offer, he could release negative emotions in his note and therefore would be less likely to express such emotions in his decision about whether or not to actually accept the monetary offer. Indeed, that is what happened. In cases of unfair offers—when the responder was offered only 10 or 20 percent of the available cash—responders expressed negative emotions in their notes to the proposers. More critical to the hypothesis, however, was the discovery that responders who wrote notes were more likely to accept unfair offers than when notes were not permitted. Given the chance to express negative emotions in a less costly way—that is, accept the unfair offer, but write a nasty note—responders chose this option over the more costly negative emotional response of rejecting the unfair offer and so receiving nothing.

Cooperate—or Else!

Positive emotions also seem to come into play in social economic interactions. In particular, emotions associated with reward processing can affect decision making. James Rilling, Ph.D., and his colleagues at Emory University had women play the Prisoner's Dilemma game either with one another or against a computer.

In this game, each player alone does better to cheat than to cooperate, but the combined payoff for mutual cooperation is greater than when both players cheat. If Sally cooperates, but Jane doesn't, Jane avoids any of the costs of cooperation but can parasitize Sally's goodness. In this case, Jane receives a higher payoff than Sally and hence is tempted to cheat. But Sally should employ the same logic, as she would do better if Jane pays the costs and she (Sally) sits back and receives the benefit of Jane's action. The catch is that when Sally and Jane both employ this logic—when they both cheat—they do worse than when they both cooperate. The dilemma in the Prisoner's Dilemma game is how to salvage cooperation when the temptation to cheat is ever present. When people play Prisoner's Dilemma many times, as they did in the Rilling experiment, one economic strategy is to be conditionally cooperative: cooperate when a partner cooperates, but not when she cheats. This strategy, called "tit for tat," allows for the relatively high payoff for mutual cooperation and minimizes the number of times players get nothing.

The Prisoner's Dilemma game again highlights the interdisciplinary nature of neuroeconomics because the same game is frequently used to analyze the biological evolution of cooperation. In the evolutionary version of the game, which uses computer simulation, a person's fitness is increased or decreased depending on whether he cooperates; the results are tracked over many simulated computer generations of players. Tit for tat works well in this version of the game; thus, the results from the study of Prisoner's Dilemma in neuroeconomics may shed light on the evolution of human cooperation as well.

Rilling and his team found that although the highest monetary reward in this game ($3) was obtained when a player cheated and her partner cooperated, the payoff that was most emotionally rewarding was mutual cooperation (which yielded $2 to each player). Subjects said that they found mutual cooperation the most rewarding outcome, and the mutual cooperation payoff caused the greatest activation of brain sections associated with reward processing (the ventromedial/orbitofrontal cortex, the anterior cingulated cortex and the nucleus accumbens).

Neuroeconomists have examined emotions and rational behavior with respect to cooperation but also in terms of what sections of the brain are active when a player punishes others who fail to act cooperatively. Punishing people who violate social norms interests anthropologists and evolutionary biologists who want to understand how social norms evolved before modern legal codes were in place. In particular, such punishment often entails a cost to the people who mete out the punishment and a benefit to others not even involved in the interaction. This is difficult to explain from a "selfish gene" evolutionary perspective, which posits a sort of cost-benefit ledger, favoring those traits that have high benefits and low costs to the person. Dominique de Quervain, Ph.D., and his colleagues at the University of Zurich hypothesized that one mechanism involved in maintaining this sort of punishment lies in the pleasure that people derive from enforcing social norms.

To test this hypothesis, de Quervain and his colleagues used PET scanning to watch the reactions of pairs playing what is called the Trust Game. In this game, both of the players, who are not allowed to communicate with each other, begin with 10 units of money (called monetary units, or MUs, in the experiment). Player A begins the game by deciding whether to give his 10 MUs to Player B or keep them for himself. If A opts to give B his money, then the investigator quadruples the gift to 40 MUs, so that B now has 50 MUs and A is broke. Then B is given a choice. He can send half of his MUs back to A or keep everything for himself. So if A correctly trusts B to send the money back, they each end up with 25 MUs; whereas if A opts to give nothing, they both end up with only their original 10 MUs.

De Quervain hypothesized that if A trusts B to play fairly and give him the money but B turns around and keeps it, A should view this as a violation of trust and social norms and therefore want to punish B. To allow for this possibility, one minute after B makes his decision, A can opt to punish B by revoking up to 40 MUs. (In one variant, A pays a price for doing so, but in others he does not.) If A trusts B, and B then violates that trust, A indeed punishes B—even if it is costly for A to do so. More to the point, subjects said that they enjoyed punishing players who violated their trust, and PET analysis showed that one section of the brain associated with reward (the dorsal striatum) was most active when A undertook his act of retribution.

Clearly, players took pleasure in punishing cheaters. They not only wanted to inflict some sort of economic damage on cheaters but also enjoyed evening the score. Results of the experiments also suggested that the more intense the punishment doled out to cheaters, the more active is the dorsal striatum of the players exacting retribution. In one further twist, A was sometimes told that B's decision was determined by a random device, so the decision was out of B's hands. In that situation, when B did not send money back to A, A did not view B's action as a violation of trust and did not respond by punishing B.

This version of the experiment shows that, when subjects dole out punishment to partners who violate social norms, they do not rely on those sections of their brain associated with calculated actions; they respond out of emotion. In a refinement of the Trust Game, Read Montague, Ph.D., and his colleagues at Baylor University examined what happened when a pair of subjects played 10 times, while the brains of both subjects were monitored. Montague and his colleagues found that trust between subjects developed over time and that one brain area in particular—the head of the caudate nucleus, which is associated with prioritizing and ranking rewards—was active when subjects were determining the fairness of a partner's action and how to repay that action with an act of trust. The question then becomes why humans derive pleasure from punishing cheaters in order to enforce social norms, even when such enforcement is costly. An evolutionary biologist might ask: What are the hidden benefits for survival that people may gain, or that our long-ago ancestors might have gained, from acting as enforcer? The answer, which is explained shortly, may center on reputation.

Some Reasonable Emotions

Neuroeconomics seems to demonstrate that the brain is hardwired to handle some economic problems through emotion rather than number crunching. Experiments have shown that specific areas of the brain, known to process specific emotions, are activated during various economic decision-making scenarios. No one suggests that there is a "fairness center" in the brain; in fact, different brain areas are active when the question of fairness arises in social versus asocial situations. But human brains do seem to have evolved to respond with special emotional vehemence to social cheating.

I am not suggesting that humans have evolved a specific emotional response to each economic situation. Instead, people have probably evolved emotional responses that, on average, work well in social situations similar to those our species faced in evolutionary history. In their experiments, neuroeconomists re-create such social situations and observe what parts of our brain are involved in our responses to them. Almost certainly, by the way, our responses occur even when we are not conscious of how our emotions are affecting our social decisions. Michael Raleigh,

Ph.D., has found that baboons whose behavior is friendly and cooperative have lots of serotonin-2 receptors in the ventromedial frontal lobes of their brains. Uncooperative, aggressive baboons do not. This may offer a biological explanation for the social behavior of baboons, but that does not mean that baboons are conscious of how their emotions affect their behavior. Likewise, our human brains may be responding to emotions when we make economic decisions about cooperation, cheating and punishment even if we are not aware of those emotions and how they influence us.

Tweaking the Rational-Man Model

Society is left with a paradox. Damasio's work and other studies of patients with brain disorders suggest that emotions are necessary for rational behavior, yet at least some work in neuroeconomics suggests that emotions often seem to act to produce irrational results, such as turning down an unfair offer at the cost of receiving nothing at all. What are we to make of this apparent contradiction?

One could argue that the contradiction is, indeed, apparent, not real. Emotions are clearly necessary for normal social interactions, but not because in themselves they always lead to rational results—if by rational we mean maximizing some simple gain such as immediate money. But acting rationally even in one-on-one social economic interactions is more complex than that. How I interact with you has implications not only for that interaction but also for my standing in a community, my reputation and my sense of self-worth—all of which have long-term economic consequences, as well as other consequences. Perhaps that is where emotions come into play, because they facilitate behaving in such a way as to promote reputation, self-worth and other long-term social assets that are also economic assets.

Do discoveries from neuroeconomics suggest that the rational-man perspective on human behavior be abandoned? I don't think so, but the rational-man model needs to be modified by what has been learned. With respect to primarily asocial economic interactions, the rational-man model works well; people behave as predicted and, what's more, everyone expects them to behave that way. It is when economic interactions are cast in a social context that this model needs to be modified. Scientists need to build emotion and reputation into the model of rational man.

Emotions do not necessarily cause people to behave irrationally in social interactions; this is true even in the studies reviewed here. Instead, once people recognize that their reputation has long-term economic consequences, scientists see that emotional responses, although they may reduce immediate gains, may foster higher overall long-term gains. If you and I are in some economic interaction that mimics the Ultimatum Game and I turn down an unfair offer that you make, I walk away with nothing. If that were the only implication of my decision, then I clearly acted in an irrational manner. But if, as is likely in real-life interactions, by turning down your offer I send a message to others that I will not accept unfair offers, then the gain in reputation may more than make up for any short-term loss.

Emotions may represent an acquired integration of the exceedingly complex calculations that are involved in such decisions. Operating automatically and almost instantaneously, they produce a response to which the brain-damaged Gage and Elliot had lost access and that they could not replace with any amount of explicit calculation of considerations and options.

Reputation

Studies of neuroeconomics and human social dynamics are new, but theories about reputation have been around for decades. Economists Robert Frank, Ph.D.; and Thomas Schelling, Ph.D.; political scientist Robert Axelrod, Ph.D.; and evolutionary biologist Randolph Nesse, M.D., have long argued that reputation—and, in particular, a reputation for keeping one's commitments—is essential to human social dynamics. (For an excellent recent synopsis, see Nesse's "Evolution and the Capacity for Commitment," Russell Sage Publishing, 2001.) In Frank's terminology, humans have evolved "passions within reason." He means that the pursuit of self-interest in a socially complex landscape requires the use of emotions to guide people along, and they may lead to actions we cannot explain without reference to reputation or standing in a community. In his masterly book Passions Within Reason: The Strategic Role of the Emotions (Norton Publishing, 1988) Frank offers this hypothetical scenario:

Jones has a $200 leather briefcase that Smith covets. If Smith steals it, Jones must decide whether to press charges. If Jones does, he will have to go to court. He will get his briefcase back and Smith will spend 60 days in jail, but the day in court will cost Jones $300. Since this is more than the briefcase is worth, it would clearly not be in his material interest to press charges. ... Thus, if Smith knows Jones is a purely rational, self-interested person, he is free to steal the briefcase with impunity. Jones may threaten to press charges, but this threat would be empty. But now suppose that Jones is not a *pure* rationalist; that if Smith steals his suit-case, he will become outraged, and think nothing of losing a day's earn-ings, or even a week's, in order to see justice done. If Smith knows that Jones will be driven by emotion, not reason, he will let the briefcase be. If people *expect* us to respond irrationally to the theft of our property, we will seldom *need* to, because it will not be in their interests to steal it.

Reputation matters, which is not to say that the drive for reputation always yields the sorts of actions that society approves. The Hatfields and the McCoys, Frank notes, killed each other for 40 years, apparently motivated by concern for defending reputation. Either family could have stopped the cycle of violence at any time simply by not retaliating against the latest murder, but that would have led to the reputation that one could kill their clan members without fear of reprisal, and, from their perspective, that would not do.

Mathematical models can help us understand how reputation may have evolved as a survival advantage. Consider a model of the Ultimatum Game developed by Martin Nowak, Ph.D., and his colleagues. They created a computer simulation in which cyber-players find themselves in the role of either proposer or responder. Proposers offer some proportion of the money available, whereas responders have a minimum acceptable offer. Some proposers make fair offers and some do not; some responders have high acceptance thresholds and others have low acceptance thresholds. How would each fare over the long term?

Nowak's simulation assigns strategies that define how each cyberplayer acts in both the proposer and the responder roles: For example, one strategy might be to offer low when in the role of proposer but accept only high offers when in the role of responder. The computer simulation then keeps track of the success of different strategies for some number of "cybergenerations." Strategies that do well increase their "evolutionary fitness" and their representation in the next generation, thus mimicking the process by which natural selection favors the spread through a species of behavior advantageous to survival and reproduction.

Nowak and his team found that, when no reputation was built into their game, the most favorable evolutionary solution was to offer low as a proposer and accept low as a responder. That is, the evolutionary solution matched the solution from a traditional rational-man economic perspective. Once reputation was built into this model by giving a player information about his partner's previous behavior, however, a different outcome emerged. Now the evolutionary solution was to make a fair offer (half the resource) and to accept only fair offers, which is what people practice in modern neuroeconomic studies. In other words, people in the Ultimatum Game behave as the model predicts, but only when reputation is taken into account.

Neuroeconomics is a young field, but its potential seems great. With the power of fMRI, PET, and computer simulations and well-grounded economists, social psychologists, neurobiologists and evolutionary biologists, the prospects for a more fundamental understanding of human social dynamics are better than ever.

CHAPTER 14

The Quest for Longer Life

Mortal Coil: A Short History of Living Longer
by David Haycock
(Yale University Press, 2008; 320 pages, $30)

Reviewed by Mark P. Mattson, Ph.D.

Mark P. Mattson, Ph.D., is chief of the Laboratory of
Neurosciences at the National Institute on Aging and
professor of neuroscience at Johns Hopkins University. He
is editor in chief of NeuroMolecular Medicine and Ageing
Research Reviews, and part of the editorial team of the journal
Neurobiology of Aging, the Journal of Neurochemistry and
the Journal of Neuroscience Research. His research specializes
in the molecular and cellular processes of neurodegenerative
diseases and age-related neurological disorders.

HUMAN BEINGS HAVE LONG BEEN FASCINATED and in many cases obsessed with the possibility of living forever, or even the less ambitious goal of living longer before experiencing disability and disease. Tales of individuals with exceptional longevity—including Methuselah and other characters in the Bible—have been passed from generation to generation throughout the history of modern man, despite a lack of substantive evidence. However, with the advent of actuarial and scientific methods it has become clear that, as with other biological variables, the life spans of individuals in populations throughout the world follow a bell curve, with the maximum attainable life span of approximately 120 years remaining largely unchanged for thousands of years.

In his new book, Mortal Coil, David Haycock, a historian and an authority on the history of medicine, provides a riveting account of the past four centuries of humans' search for the explanation of their mortality and the possibility of achieving immortality. The book integrates religious, philosophical and scientific considerations of mortality, using well-researched accounts of the lives and contributions of key thinkers as well as of charlatans who have shaped our views of aging. Drawing upon his training as a historian and his knowledge of medicine, Haycock provides a unique view of a fundamental aspect of human existence, written in an easy-to-read yet thought-provoking manner—albeit with room for further insight about the brain.

The book begins in the early 1600s with the life and intellectual pursuits of Sir Francis Bacon, who made several major contributions that advanced knowledge of the causes of aging and the prospect of life-span extension. In his essays on this subject, Bacon describes how the aging process is unequal—some parts of the body age relatively fast and are not repaired (the arteries and bones, for example), whereas other parts, such as the skin and the blood, can be restored during aging. In a time when the scientific method was not yet developed and applied to the problem of aging and mortality, he emphasized the importance of acquiring and assimilating knowledge, as exemplified by his famous quote, "Knowledge is power."

Interestingly, Bacon recommended a diet that included plants known to contain poisonous substances, such as hemlock, mandrake and night-shade. Recently, accumulated scientific evidence has suggested that the health benefits of some of the chemicals in plants do indeed result from their noxious properties; in a process called "hormesis," low amounts of such potentially toxic substances induce a mild adaptive stress response in cells and organs, which results in increased resistance to disease. As described by Haycock, it was also in the 17th century that Rene Descartes emphasized the importance of dietary moderation, exercise and "peace and tranquility." This section of the book provides an interesting and well-conceived account of how people viewed mortality from religious and philosophical perspectives at a time when the biological underpinnings of aging were completely unknown.

The era of alchemy and the search for the "elixir of life," still in the 1600s, is the topic of the next section of Mortal Coil. Haycock describes the intermingling of the beginnings of medicine with the promulgation of anecdotal cases and outright witchcraft and quackery that were fueled by the hopes and fears of the sick and the aged. Claims of dramatic cures through treatments ranging from bloodletting to herbs to gonadal extracts were rampant. Despite the prevalence of false claims during this time period, however, valuable information and concepts were generated. For example, the Swiss alchemist and physician Paracelsus developed the idea that medicines are poisons that at low doses have beneficial effects. In England, there was a growing interest in the prolongation of life among members of the Royal Society, and physicians commonly employed drastic measures to prolong the lives of people on their deathbeds. Robert Boyle espoused the "corpuscular theory"—the principle that life could be prolonged by replacing body fluids and cells in old individuals with fluids and cells from younger individuals.

Haycock describes the 18th century as the age of reason and optimism for the future of medicine, health and longevity. Claims of exceptional longevity continued—St. Germain claimed to be 300 years old, and even prominent physicians such as the Scottish doctor George Cheyne suggested that it was possible to live beyond 200 years by adopting a strict Spartan-like lifestyle. But this hope among intellectuals and medical professionals was in stark contrast to the reality of the times, for the average life span was actually decreasing; for example, average life expectancy in England in 1726 was 25 years.

One lesson of the 1700s, however, was that with optimism come energy and motivation, and, indeed, the belief that immortality might be eventually achieved through advances in science and medicine became an impetus for rigorous research in these fields. Even Napoleon Bonaparte expressed the opinion that science would find a way to prolong life indefinitely. And on the other side of the Atlantic Ocean in sprouting America, Benjamin Franklin was fascinated with the process of aging and the possibility of slowing it through dietary modifications—he found that his own health benefited by abstaining from meat. Also important among the conceptual advances of the 1700s were the idea of the "power of the mind" forwarded by William Godwin and the appreciation of the role of inheritance as a determinant of one's life span as voiced by the German physician Christof Hufeland. This section of the book touches on the influence of the brain on health and longevity, but could have benefited from specific examples of how chronic stress shortens life and engagement in intellectual and leisure activities promotes health. Interestingly, many of the environmental factors that were anecdotally linked to increased longevity in the 16th through the 18th centuries exert their beneficial effects by modifying neurotransmitter systems in the brain. For example, by increasing signaling by the neurotransmitters serotonin and glutamate, exercise and engagement in intellectual activities increase the production of neurotrophic factors, proteins that promote the growth and survival of nerve cells and thereby protect against disease.

Rigorous investigations of longevity and the aging process, in which reliable data were systematically collected, began and then grew exponentially during the 19th century. Data collected by the life insurance industry based on substantiated birth and death records definitively refuted claims that individuals were living beyond 120 years of age. Thomas Malthus' analysis of population growth demonstrated the importance of competition for available resources in setting limits on survival, and predicted adverse consequences of continued exponential growth of the human population. At the same time, the meticulous research of the naturalists Alfred Wallace and Charles Darwin led to the realization that humans evolved over millions of years by the process of natural selection. These kinds of fully validated discoveries clearly suggested that longevity is largely predetermined by evolutionary history and that, accordingly, the death of the individual is important for the survival of the species. Elsewhere, Benjamin Gompertz developed an equation that describes the survival curves for populations of any animal from which two key values can be obtained: the average life span and the maximal life span. Coincident advances in medicine by famous scientists such as Louis Pasteur led to reduction in deaths caused by infectious agents, which resulted in increases in the average life span—but without an increase in the maximum life span, which remained at 100 to120 years.

Haycock concludes Mortal Coil by summarizing the many advances in aging research and in understanding the molecular and cellular basis of aging and age-related disease. This large body of knowledge, the vast majority of which has been obtained within the past hundred years, was bolstered by numerous major discoveries in the field of biology, including what cells are and how they divide and function; the structure and mechanism of replication of DNA; how proteins, which are made from amino acids, control the structure and function of cells; and the nature of oxygen free radicals and how they damage cells.

Though not overtly stated by Haycock, the title for his book presumably derives from the Shakespearean use of the phrase: Hamlet, in his famous soliloquy, ponders what happens once we have "shuffled off this mortal coil"—the troubles of life and the suffering in the world. But the title also invokes the double helix "coil" of DNA, inherited from our parents, which controls our life and also dooms us to death. Accordingly, included in the last section of the book are sketches of the work of several prominent scientists whose discoveries support the importance of the "mortal coil," among them Thomas Kirkwood, who developed the "disposable soma theory of aging." This theory focuses on the concept that evolution protects the germ cells while discarding the body through its programmed death. The discovery that dividing cells in the body, such as those in the skin, are capable of only a limited number of cell divisions, proportional to the life span of the species, strengthened the case for a genetically programmed life span. Haycock touches on key points concerning the mechanisms of aging, but leaves it to the reader to integrate this information with that covered in the previous sections of the book. The book would have been enhanced by the inclusion of examples of predictions made in the preceding centuries and whether they were or were not supported by emerging scientific findings.

Some of these examples are brain-related. Many centuries before the discoveries of nerve cells and neurotransmitters, it was recognized that thought, mood and behavior could be affected by substances present in certain plants and animals. A culture of using traditional medicines that were mainly components (roots, leaves, bark, etc.) or extracts of plants became pervasive in societies throughout the world. In an increasing number of cases, the efficacy of such traditional medicines is being validated in controlled studies, and the specific chemicals responsible for the medicinal actions are being identified. In several cases, the active chemicals are actually toxins that, at the low doses consumed, activate adaptive cellular stress response pathways in neurons and other cells. The result is that the cells respond to the mild stress induced by the phytochemicals by increasing their ability to cope with more severe stress and resist disease. It would have been more informative if Haycock had included a description of the brain circuitry and neurotransmitters that control mood (serotonin, norepinephrine and dopamine) and how drugs used to treat depressed or otherwise abnormal mood act on these neurotransmitter systems. In modern societies, mood-altering drugs, particularly antidepressants, are becoming widely used. It will be of considerable interest to know how views of longevity and life span are affected by these drugs.

Also missing from the book is discussion of the roles of the brain in the contemplation of mortality and in efforts to delay death. From an evolutionary perspective, survival is paramount, and the brain is of fundamental importance in the critical decision-making processes that determine one's fate. As modern societies evolved, survival depended less on avoiding sudden violent death and more upon avoiding and properly treating diseases. Assimilation of scientific data and the development of drugs in pharmaceutical companies and universities required the coordinated efforts of individual scientists and physicians—a dramatic example of altruism. Higher cognitive functions were essential for these efforts to extend health span (the number of years lived in good health) and, with it, average life span. Because of the brain's structural and chemical complexity, and the prevalence of psychiatric and neurological disorders such as schizophrenia, depression, epilepsy, Alzheimer's disease and Parkinson's disease, the brain was itself a major target for drug development. Together with basic research in animals and human subjects, the various neurotransmitters that control all behaviors and bodily functions were identified, and "neurochemical maps" of the nerve cell circuitry of the brain were established.

While great strides have been made in understanding what happens to the body's molecules and cells during aging, the general "formula" for living longer and healthier that was first appreciated four centuries ago remains largely unchanged: eat in moderation, exercise regularly, keep the mind engaged in challenging intellectual pursuits and avoid chronic stress. In 2008, it is realistic to expect further incremental advances that increase average life span and extend health span but, as suggested by the history elegantly chronicled by Haycock in Mortal Coil, the prospect of major increases in maximum life span remains remote.

CHAPTER 15

The Brain,
from Atom to Soul

A Portrait of the Brain
by Adam Zeman
(Yale University Press, 2008; 256 pages, $27.50)

Reviewed by Lewis P. Rowland, M.D.

Lewis P. Rowland, M.D., is professor of neurology at Columbia University. His research has increased our understanding of neuromuscular diseases and age-related neurodegenerative diseases, especially ALS. He is the chief editor of Neurology Today, a biweekly newspaper publication of the American Academy of Neurology, and editor of a number of medical textbooks, including the last five editions of Merritt's Neurology.

WE ARE BLESSED THESE DAYS with an abundance of physician-writers who are interested in neurology. At the top of the list, of course, is Oliver Sacks. Now he and the others are joined by one more neurologist, Adam Zeman, a professor of cognitive and behavioral neurology at the new Peninsula Medical School in Plymouth, England. The school graduated its first class in 2007.

Zeman writes smoothly and with a flair for the well-turned phrase. Each chapter is devoted to a different topic and comprises three elements: a particular aspect of neuroscience, a related explanation of a particular neurological disease, and a case report that humanizes that disorder. The author knows the patients and their symptoms well because they come from his own practice. In discussing the relevant science, he describes the history of our concepts about the disease because, he writes, "the history of a science is the science."

Zeman employs another unique idea as well: the chapters progress sequentially from the most basic to increasingly more complex units of life. Chapter 1, "Atom," describes the importance of oxygen for human life and exemplifies the author's fearlessness; he admits a medical error in the opening chapter. The patient, a 38-year-old woman, was thought to have chronic fatigue syndrome until her lips turned blue and she lapsed into a coma. Muscle biopsy disclosed a rare congenital disorder called multicore myopathy. With assisted ventilation through the night, she emerged from the coma, nocturnal sleep was restored, and her excessive daytime fatigue was ameliorated. Zeman concludes: "Instead of nervously concealing them, we should examine, even celebrate, our failures and mistakes. Rather than being negligent or shameful, as a rule, they are a fact of life, a plain reflection of human imperfection."

Chapter 2, "Gene," deals with diseases characterized by involuntary movements, some of which are inherited. That leads to analysis of DNA by way of acanthocytes, from a Greek word meaning "thorn." These prickly-looking red blood cells appear in only a few diseases. In this case, recognition of the misshapen cells led to a diagnosis of McLeod syndrome rather than Huntington disease, the equally malignant cause of chorea. "Chorea" encompasses the random jerking of limbs and derives from the

Greek word for dancing (whence, also, comes "chorus"). In the context of this patient, "Charley," Zeman celebrates the achievements of molecular genetics: "Most of the genes that cause acanthocytosis have been lassoed in the past ten years. Around 60 genes causing inherited forms of inco-ordination and unsteadiness and a dozen causing inherited Parkinson's disease have been described at a rapidly accelerating pace." But then comes the jolt: "Delivering really effective treatment for these disorders … is, so far, an unrealized ambition. For Charley, the discovery of the cause of his predicament provided cold comfort. I was unable to rescue him from his fate. His last few years were spent in psychiatric custody." Zeman's comment is a constant source of general concern: discovering genes has been a major advance for medicine but, lamentably, treatment for these diseases is largely unavailable. Gene therapy has commenced, but it has a long way to go.

Chapter 3, "Protein," starts with a description of scrapie, a disease of sheep. Back in 1936, two French veterinarians proved that if brain tissue from an infected animal was injected into the brain of a normal animal, the disease appeared a year or two later. The first Nobel Prize for research on this kind of rare disease was awarded to Carlton Gajdusek, who confirmed and elaborated the transmissibility of Creutzfeldt-Jakob disease (CJD); when brain samples from a victim of the killer disease were injected into a monkey, the recipient later developed the telltale symptoms. So came the era of "slow virus infection," "transmissible spongiform encephal-opathy," and, ultimately, "prion disease." The case example here is one of Creutzfeldt-Jakob disease, with appropriate attention also to the other diseases in the category as well as the causal self-replicating proteins called prions. That word was invented by Nobelist Stanley Prusiner, who won the second Nobel for work on a disease as rare as CJD; he described the infectious agent as one that replicates but has no DNA, a wondrous excep-tion to conventional laws of genetics. Prusiner called the agent "proteina-ceous infectious material," from which came the name "prion" (which he pronounced "pree-on").

Continuing up the ladder of cellular complexity, we reach "Organelle." The subtitle of this chapter is "Metamorphoses," which describes the bacterial origin of human mitochondria millions of years ago and the current diseases of these intracellular structures. Salvatore DiMauro, Michio Hirano and Eric Schon have written a monograph in this field, which they call "mitochondrial medicine." The term encompasses several neurological syndromes, including Kearns-Sayre syndrome, mitochondrial encephalomyopathy with lactic acidosis and stroke-like episodes (MELAS) and others. Non-neurological manifestations include diabetes mellitus and fatty deposits or "lipomas" under the skin. Perhaps more significant, mitochondrial malfunction is considered important in the many age-related degenerative diseases of the brain and nervous system. Here, Zeman explains the basics of the maternal inheritance of mitochondrial DNA and how it has become a boon to anthropology.

Chapter 5, "Neuron," describes the debate about synapses between two giants in the history of neuroscience, Ramón y Cajal and Camillo Golgi. They shared the first Nobel Prize, in 1906, but their enmity was evident in the lectures they presented at the award ceremony. Golgi developed a stain that is still used to study the ramifications of neurons in the brain, but he concluded, in Zeman's words, that the cellular projections "unite to form a fused network of fibres." The chapter explicates the role of ions in neural transmission, and the case history describes a young man with epilepsy.

Chapters 6 through 8 take us from "Synapse" to "Neural Networks" and then "Lobe," and from narcolepsy to déjà vu and on to the emergence of creativity as dementia becomes more and more severe.

Chapter 9, "Psyche," focuses on somatoform syndromes, which are psychological disorders marked by physical symptoms for which no physiological explanation is found. Here, Zeman is brave enough to use the word "hysteria," a term that has been set aside by many writers and psychiatrists as sexist. The word "hysterical" comes from a word in both Latin and Greek that means "womb," or "uterus." The word was originally used to describe symptoms without evident cause, considered peculiar to women and attributed to dysfunction of the uterus, but we now recognize the condition in men as well as women.

The last chapter, "Soul," would be difficult for anyone—neurologist, philosopher or theologian—to tackle. Zeman writes: "The words we use when we think about the mind are quite different. 'Mind', 'soul', [and] 'consciousness' are not scientific terms, and lack technical definition. Our understanding of them is powerfully influenced by religious and philosophical traditions. So there is an inevitable risk of a disconnection here between scientific enquiry and everyday thought." Neurosurgeon Wilder Penfield demonstrated by brain stimulation in the course of epilepsy surgery that thoughts arise in the brain. Yet even he became a dualist, describing a soul separate from the flesh. Zeman is correct in ascribing the concept of a soul to the worldview of most people, but this concept is rejected by atheist philosophers, including the popular science writer Richard Dawkins and the literary critic Christopher Hitchens.

Neuroscientists, neurologists, psychiatrists and general readers should all enjoy and benefit from reading this comprehensive yet succinct book, as Zeman shows how interest in a clinical condition can lead to science, human experience—and fine literature.

CHAPTER 16

Memoirs About Memory:
Too Much Versus Too Little

Can't Remember What I Forgot
by Sue Halpern
(Harmony, 2008; 272 pages, $24)

The Woman Who Can't Forget
By Jill Price with Bart Davis
(Free Press, 2008; 272 pages, $26)

Reviewed by Suzanne Corkin, Ph.D.

Suzanne Corkin, Ph.D., directs the Behavioral Neuroscience Laboratory in MIT's Department of Brain and Cognitive Sciences. She and her colleagues address questions concerning the cognitive and neural basis of learning and memory in humans, using tests that assess specific cognitive processes, structural magnetic resonance imaging (MRI), functional MRI (fMRI), magnetoencephalography (MEG) and genotyping. Research participants include patients with global amnesia, Alzheimer's disease and Parkinson's disease, as well as young and older individuals without neurological disease.

JILL PRICE'S The Woman Who Can't Forget and Sue Halpern's Can't Remember What I Forgot: The Good News from the Front Lines of Memory Research are two memoirs that describe opposite sides of the memory chip. Price has an extraordinarily good memory, but her recollections are so frequent and vivid that they disrupt her life. Halpern, though labeled "normal," is anxious about the possibility, and maybe inevitability, of losing her memory. For Price, forgetting is a boon; for Halpern, it portends cognitive decline with aging.

Price recounts her personal saga as a patient in search of answers about her uncanny ability to remember specific dates and events without even trying. Halpern, whose 1992 Migrations to Solitude was a New York Times notable book, gives a detailed account of her visits to clinics and laboratories to ask the experts: what is memory? How does it break down in disease? How can we fix it? Each author intends to leave the reader with an optimistic message, but only Price succeeds: in closing, she expresses hope for her future and for the possibility that her case study will benefit other patients. In contrast, by the time I finished Halpern's book, I was convinced that the subtitle should have been The Bad News from the Front Lines of Memory Research.

Too Much Memory

Price begins her story: "I know very well how tyrannical the memory can be. I have the first diagnosed case of a memory condition that the scientists who have studied me termed hyperthymestic syndrome—the continuous, automatic autobiographical recall of every day of my life from when I was age fourteen on." The scientists to whom Price refers are James McGaugh, Larry Cahill and Elizabeth Parker at the University of California, Irvine. They spent 5 years studying Price in their laboratory and summarized their findings in a fascinating article in *Neurocase* (2006).[1] (In their paper, they gave her the initials A.J. to protect her identity; at the time, she had not gone public.) They began by describing Price's history in some detail and then gave examples of her extraordinary autobiographical memory.

During Price's visits to their laboratory, they quizzed her by giving her a date and asking her what had happened on that day. As rigorous scientists, they did not assume unquestioningly that all of her recollections were correct. Rather, they verified her responses against extensive diary entries and information from her mother, calendars and the media. Price's responses were highly reliable. For example, when asked about July 1, 1986, she replied, "I see it all, that day, that month, that summer. Tuesday. Went with (friend's name) to (restaurant name)." The scientists later confirmed that the day of the week was indeed a Tuesday and that the event corresponded to a diary entry. Her memory for public events was equally amazing; when asked what happened on June 6, 1978, she correctly stated that Proposition 13 passed in California. And when given the date November 4, 1979, she knew that the U.S. Embassy in Iran was invaded on that day. Clearly, her ability goes far beyond that of most.

Price's recollections are profoundly personal. An especially poignant example concerns her mother's overriding concern with Price's weight. She writes, "I put on weight, as so many people do when they go to college, and when I went home my mother harped about that. All the memories through the years of her nagging me about my weight were triggered, and they began to haunt me again relentlessly. Maybe that should have made me thrilled about being off on my own, but I felt nothing like that" (page 150). Although the book at times smacks of a TV serial, I did feel the emotional torment that Price experienced as a result of her out-of-control memory.

Since the 2006 publication of the scientific paper by the UC-Irvine scientists, other individuals with hyperthymestic syndrome have surfaced. One of them, Brad Williams, is a 51-year-old radio anchor who constantly amazes his family, colleagues, and the public with his memory prowess. He appeared on "Jeopardy" and just missed being champion. He also agreed to a contest with Google to see who could answer 20 questions about news events faster—and won. The public will soon have the opportunity to marvel at his lifetime of memory feats because his brother, Eric, a screenwriter, is producing a documentary titled "Unforgettable" (see the trailer on YouTube). Undoubtedly, more such individuals will emerge in the future, each with a fascinating story to share.

Price and Williams, and the neuroscientists who have studied them, are baffled about how they perform memory feats that most other people cannot. We do know that learning and memory take place over time, and in the laboratory we can examine the stages of memory formation. A current framework for studying memory in humans tries to isolate encoding, storage and retrieval processes, but memory is really not that simple. In addition, one would like to know whether imaging their brains would reveal any structural, connectional or functional differences from the brains of people with ordinary memory. Such work is in progress at Irvine, and Price notes in her book, "My brain scans show that my brain has some structural features that are a great deal different from the norm" (page 95). To complete the picture, a thorough personality evaluation would be useful to see whether particular traits may have nourished Price's hyperthymesia. We can count on McGaugh, Cahill and their colleagues to ask the right questions and tell us what these extraordinary minds teach us about the organization of human memory, and also how the brain's process of rewiring based on experience can go awry.

Too Little Memory

Sue Halpern's father had vascular dementia and died at age 77. His profound memory impairment and its devastating effect on his last years seized her attention. She wondered about the underpinnings of his sickness, prompting a series of questions about how memory decline is diagnosed and treated. Glimpses of the clinical science and basic science that she caught from the popular literature fascinated her, and she decided that she wanted to know more. To this end, she embarked on a quest for knowledge about memory. She enrolled as a healthy volunteer in several clinical studies and became one of the "worried well." Appropriately, the first chapter of the book is titled, "Anxious."

Halpern's book is tirelessly researched. She interviewed more than 50 clinicians and scientists over a period of several years. She chose to undergo a structural magnetic resonance imaging (MRI) scan, a functional

MRI (fMRI) scan a positron emission tomography (PET) scan, a single photon emission computed tomography (SPECT) scan and neuropsychological testing. In addition, she learned about various basic science investigations in animals and humans that aim to uncover the biology of aging and age-related disease. Halpern describes diseases, neuroimaging methods and cognitive tests in language that the layperson can understand. She writes lucidly and tries to show a healthy skepticism about promised breakthroughs that never materialized. Still, I fault the book on two counts: the subtitle, "The Good News from the Front Lines of Memory Research," which gives readers expectations that are not realized, and the inaccurate reporting of many scientific facts (examples to follow).

The Bad News

Reading this book, I wondered when I would get to the "good news." Finally, in chapter 8, I got a morsel. Halpern writes, "Exercise alone appeared to improve cognition" (page 186) and "memory improved in humans who exercised. That was now known" (page 205). This tidbit of encouragement was juxtaposed with an abundance of negative messages, such as "Almost every drug in development fails" (page 138) and, quoting Steven Ferris, "There's nothing that has been proven so far in an FDA-regulated quality sense to improve memory in normal brain aging. In the end, the proof will be made in the human pudding, and it hasn't been made yet" (page 180). And last but not least: "[If the neurologist] made a diagnosis of Alzheimer's, the best he could offer was a prescription for drugs that did not work very well, and a referral to social services, and good wishes" (pages 227–228). And so the book ends, with the final bit of bad news from the front lines of memory research.

Inaccurate Reporting

Our society needs books written by non-scientists that don't turn off the general reader by sounding overly "scientific" but nevertheless transmit up-to-date research findings. Halpern's book is an attempt at that kind of communication. Like many other trade books, however, Halpern's falls short of this goal. The reason is that she (I suspect) and many other pop authors don't immerse themselves in the scientific literature. Instead, they rely on interviews with clinicians and scientists as a substitute for going to the library. Then they write a book. The result can be a compendium of significant errors and a slippery use of facts.

Halpern writes at length about "breakthrough" treatments that amount to what strikes me as snake oil. She highlights neuroscientists with panaceas and panacea-based research programs. These individuals take meager scientific findings and make claims for them that are not supported by the evidence. In some cases the beneficiaries are the scientists themselves, or even commercial entities such as TV channels.

My major objection is to Halpern's erroneous description of the amnesic patient H.M., whom I studied from 1962 through his death Dec. 2, 2008. According to Halpern,

The reason why so much was known about how a small ridge at the bottom of the brain, one on the left, the other on the right—the hippocampus—controls short-term memory was that in 1946 a nine-year-old boy fell off his bicycle. After the accident the boy began to experience unremitting seizures, seizures so severe that when he was sixteen, with no prospect for a decent life, his doctor removed a part of his temporal lobe, including his hippocampus, hoping to eliminate the tissue that he believed was causing the young man's body to convulse uncontrollably. It was a radical move, but the doctor was right—the operation cured the young man of his seizure disorder. But it turned out that without a hippocampus, the young man was no longer able to make new memories. He could find his way to his childhood home, but not to the house where he lived after the operation. He greeted his doctors each day as if they had never before met. Over the years, when he looked in the mirror, he saw a stranger too old to be himself; his self-image had been fixed in his brain in the days and weeks leading up to the operation. (page 53)

Let me set the record straight. The hippocampus is not located at the bottom of the brain. It is situated above the ears, toward the middle of the brain. Considerable scientific evidence indicates that the hippocampus and the surrounding cortex are critical for the establishment of long-term declarative memory (not short-term memory). In 1935 (not 1946), H.M. was knocked down by a bicycle. His minor seizures began a year later, at age 10; his major seizures began at age 16. His operation occurred at age 27 (not age 16). The removal was restricted to the medial part of the left and right temporal lobes, including the hippocampus. The operation did not cure H.M.'s seizure disorder. His seizures were reduced in frequency, but for the rest of his life he took anticonvulsant medication and had seizures.

The statement concerning H.M.'s sense of self is also inaccurate. When he looked in the mirror he was not shocked. He showed no change in facial expression, his conversation was matter-of-fact, he did not seem to be at all upset and on one occasion, when asked, "What do you think about how you look?" he replied, "I'm not a boy," illustrating his sense of humor. H.M.'s sense of self included knowledge of his ancestors, preoperative personal semantic knowledge (i.e., general information about the world), memories of his childhood that included vacations with his parents and information about a number of relatives (although he could provide only the gist of these events without any specific details about time and place). In addition, amnesia is not an all-or-nothing condition, and even H.M. from time to time had meager conscious recollections of information encountered postoperatively. For example, he knew he had a memory impairment, he could draw the floor plan of his postoperative home and he had some slight knowledge about a handful of celebrities. The interested reader can learn more about H.M. in a paper I authored in Nature Reviews Neuroscience in 2002.[2] Halpern and other popular science writers would be well advised to pay closer attention to the scientific literature.

Common Themes

What do Price and Halpern have in common? Both authors acknowledge that memory is a critical component of one's sense of self. When Price reached adolescence, her autobiographical memories became more numerous and vivid. Here's how she interprets this enhancement: "The most interesting explanation to me is that most people have more memories from this time period because it is in these years that we are generally formulating and fixing in our minds our sense of self, and memory and self are closely intertwined" (page 101). For her, the flood of memories gave rise to a negative self-image and an unhappy life. In the epilogue, she considers whether she would have been better off without her exceptional memory. Her answer is, "Despite all the pain it has caused me, if I could choose, I would keep my memory, because it's made me who I am" (page 246). Price has learned to live with, and perhaps even appreciate, her unusual persona, and she accepts the challenge of keeping it under control. Brava!

Halpern also believes that we are what we remember: "It is not just that the brain codes in terms of spatial-temporal patterns, it's that our memory is what places us, as individuals, in space and time. It's through our memory that we know where we are, and where we are going, and who we are, and who we believe others know us to be" (page 221). That said, the challenge of understanding memory persists.

Endnotes

1. COMING APART:
TRAUMA AND THE FRAGMENTATION OF THE SELF

1. J. J. Freyd, B. Klest and C. B. Allard, "Betrayal Trauma: Relationship to Physical Health, Psychological Distress, and a Written Disclosure Intervention," *Journal of Trauma and Dissociation* 6, no. 3 (2005): 83–104.

2. L. Cahill et al., "Beta-adrenergic Activation and Memory for Emotional Events," *Nature* 371 (1994): 702–704.

3. M. C. Anderson et al., "Neural Systems Underlying the Suppression of Unwanted Memories," *Science* 303, no. 5655 (2004): 232–235.

4. L. M. Williams, "Recall of Childhood Trauma: A Prospective Study of Women's Memories of Child Sexual Abuse," *Journal of Consulting Clinical Psychology* 62 (1994): 1167–1176.

5. M. A. Conway, "Cognitive Neuroscience: Repression Revisited," *Nature* 410, no. 6826 (2001): 319–320.

6. E. Geraerts et al., "The Reality of Recovered Memories: Corroborating Continuous and Discontinuous Memories of Childhood Sexual Abuse," *Psychological Science* 18, no. 7 (2007): 564–568.

7. J. D. Bremner et al., "Dissociation and Posttraumatic Stress Disorder in Vietnam Combat Veterans," *American Journal of Psychiatry* 149, no. 3 (1992): 328–332.

8. K. Ginzburg et al., "Evidence for a Dissociative Subtype of Post-traumatic Stress Disorder Among Help-Seeking Childhood Sexual Abuse Survivors," *Journal of Trauma and Dissociation* 7, no. 2 (2006): 7–27.

9. M. G. Griffin, P. A. Resick and M. B. Mechanic, "Objective Assessment of Peritraumatic Dissociation: Psychophysiological Indicators," *American Journal of Psychiatry* 154, no. 8 (1997): 1081–1088.

10. R. A. Lanius et al., "Neural Correlates of Traumatic Memories in Posttraumatic Stress Disorder: A Functional MRI Investigation," *American Journal of Psychiatry* 158, no. 11 (2001): 1920–1922.

11. R. A. Lanius et al., "Brain Activation During Script-Driven Imagery Induced Dissociative Responses in PTSD: A Functional Magnetic Resonance Imaging Investigation," *Biological Psychiatry* 52, no. 4 (2002): 305–311.

12. C. Koopman et al., "Dissociative Symptoms and Cortisol Responses to Recounting Traumatic Experiences Among Childhood Sexual Abuse Survivors with PTSD," *Journal of Trauma and Dissociation* 4, no. 4 (2003): 29–46.

13. E. Vermetten et al., "Hippocampal and Amygdalar Volumes in Dissociative Identity Disorder," *American Journal of Psychiatry* 164, no. 4 (2006): 630–636.

14. D. Spiegel, "Recognizing Traumatic Dissociation," *American Journal of Psychiatry* 163, no. 4 (2006): 566–568.

15. R. F. Pitman, "Hippocampal Diminution in PTSD: More (or Less?) Than Meets the Eye," *Hippocampus* 11, no. 2 (2001): 73–74, discussion 82–84.

16. O. T. Wolf et al., "Basal Hypothalamo-Pituitary-Adrenal Axis Activity and Corticotropin Feedback in Young and Older Men: Relationships to Magnetic Resonance Imaging-Derived Hippocampus and Cingulate Gyrus Volumes," *Neuroendocrinology* 75, no. 4 (2002): 241–249.

17. C. A. Morgan et al., "Symptoms of Dissociation in Healthy Military Populations: Why and How Do War Fighters Differ in Responses to Intense Stress?" in *Traumatic Dissociation Neurobiology and Treatment*, ed. E. Vermetten, M. J. Dorahy, and D. Spiegel, 157–179 (Washington, D.C.: American Psychiatric Publishing, 2007).

18. J. H. Krystal et al., "Toward a Cognitive Neuroscience of Dissociation and Altered Memory Functions in Post-traumatic Stress Disorder," in *Neurobiological and Clinical Consequences of Stress: From Normal Adaptation to PTSD*, ed. M. J. Griedman, D. S. Charney, and A.Y. Deutch, 239–269 (Philadelphia: Lippincott-Raven, 2005).

4. A ROAD PAVED BY REASON

1. S. D. Hollow, R. J. DeRubeis, R. Shelton, J. D. Amsterdam, R. M. Salomon, J. P. O'Reardon, M. L. Lovett, P. R. Young, K. L. Haman, R. B. Freeman and R. Gallop. 2005. Prevention of relapse following cognitive therapy vs. medications in moderate to severe depression. *Archives of General Psychiatry* 62:417–422.

2. A. C. Butler, J. E. Chapman, E. M. Forman and A. T. Beck. 2006. The empirical status of cognitive-behavioral therapy: A review of meta-analyses. *Clinical Psychology Review* 26:17–31.

3. S. G. Hofmann and J. A. J. Smits. 2008. Cognitive-behavioral therapy for adult anxiety disorders: A meta-analysis of randomized placebo-controlled trials. *Journal of Clinical Psychiatry* 69:621–32.

4. K. Goldapple, Z. Segal, C. Garson, M. Lau, P. Bieling, S. Kennedy and H. Mayberg. 2004. Modulation of cortical-limbic pathways in major depression. *Archives of General Psychiatry* 61:31–41.

5. G. Parker, K. Roy and K. Eyers. 2003. Cognitive behavior therapy for depression? Choose horses for courses. *American Journal of Psychiatry* 160:828–834.

6. P. Bolton, J. Bass, R. Neugebauer, H. Verdeli, K. F. Clougherty, P. Wickramaratne, L. Speelman, L. Ndogoni and M. Weissman. 2003. Group interpersonal psychotherapy for depression in rural Uganda. *Journal of the American Medical Association* 289:3117–3124.

7. P. Bolton, J. Bass, T. Betancourt, L. Speelman, G. Onyango, K. Clougherty, R. Neugebauer, L. Murray, and H. Verdeli. 2007. Interventions for depression symptoms among adolescent survivors of war and displacement in northern Uganda: A randomized controlled trial. *Journal of the American Medical Association* 298:519–527.

5. THE POLITICAL BRAIN

1. Editorial, "Mind Games: How not to mix science and politics," *Nature* 450 (2007): 457.

2. Jonas T. Kaplan, Joshua Freedman and Marco Iacoboni, "Us versus them: Political attitudes and party affiliation influence neural response to faces of presidential candidates," *Neuropsychologia* 45, no. 1 (November 2007): 55–64.

3. Silvia Galdi, Luciano Arcuri and Bertram Gawronski, "Automatic mental associations predict future choices of undecided decision-makers," *Science* 321 (2008): 1100–1102.

6. A WOUND OBSCURE, YET SERIOUS: CONSEQUENCES OF UNIDENTIFIED TRAUMATIC BRAIN INJURY ARE OFTEN SEVERE

1. S. Okie, "Reconstructing Lives: A Tale of Two Soldiers," *New England Journal of Medicine* 355 (2006): 2609–2615.

2. J. A. Langlois, W. Rutland-Brown and K. E. Thomas. *Traumatic Brain Injury in the United States: Emergency Department Visits, Hospitalizations and Deaths* (Atlanta: Centers for Disease Control and Prevention, National Center for Injury Prevention and Control, 2004).

3. J. F. Kraus and D. L. McArthur, "Epidemiologic Aspects of Brain Injury," *Neurologic Clinics* 14 (1996): 435–450.

4. D. M. Bernstein, "Recovery from Mild Head Injury," *Brain Injury* 13 (1999): 151–172.

5. J. M. Silver, R. Kramer, S. Greenwald and M. Weissman, "The Association Between Head Injuries and Psychiatric Disorders: Findings from the New Haven NIMH Epidemiologic Catchment Area Study," *Brain Injury* 15 (2001): 935–945.

6. T. Tanielian and L. H. Jaycox, eds., *Invisible Wounds of War: Psychological and Cognitive Injuries, Their Consequences and Services to Assist Recovery* (Santa Monica, Calif.: RAND Corporation, MG-720-CCF, 2008).

7. K. Brewer-Smyth, A. W. Burgess and J. Shults, "Physical and Sexual Abuse, Salivary Cortisol and Neurologic Correlates of Violent Criminal Behavior in Female Prison Inmates," *Biological Psychiatry* 55 (2004): 21–31.

8. M. Sarapata, D. Herrmann, T. Johnson and R. Aycock, "The Role of Head Injury in Cognitive Functioning, Emotional Adjustment and Criminal Behaviour," *Brain Injury* 12 (1998): 821–842.

9. B. Slaughter, J. R. Fann and D. Ehde, "Traumatic Brain Injury in a County Jail Population: Prevalence, Neuropsychological Functioning and Psychiatric Disorders," *Brain Injury* 17 (2003): 731–741.

10. J. S. Burg, L. M. McGuire, R. G. Burright and P. J. Donovick, "Prevalence of a Head Injury in an Outpatient Psychiatric Population," *Journal of Clinical Psychology in Medical Settings* 3 (1996): 243–251.

11. L. M. McGuire, R. G. Burright and R. Williams, "Prevalence of Traumatic Brain Injury in Psychiatric and Non-Psychiatric Patients," *Brain Injury* 12 (1998): 207–214.

12. M. Hibbard, S. Uysal, K. Kepler, J. Bogdany and J. M. Silver, "Axis I Psychopathology in Individuals with TBI," *Journal of Head Trauma Rehabilitation* 13, no. 4 (1998): 24–39.

13. L. Ewing-Cobbs, M. A. Barnes and J. M. Fletcher, "Early Brain Injury in Children: Development and Reorganization of Cognitive Function," *Developmental Neuropsychology*. 24 (2003): 669–704.

14. H. G. Taylor, "Research on Outcomes of Pediatric Traumatic Brain Injury: Current Advances and Future Directions," *Developmental Neuropsychology* 25 (2004): 199–225.

15. A. Glang, B. Todis, C. W. Thomas, D. Hood, G. Bedell and J. Cockrell, "Return to School Following Childhood TBI: Who Gets Services?" *NeuroRehabilitation*, in press.

16. E. Finkelstein, P. Corso and T. Miller, *The Incidence and Economic Burden of Injuries in the United States* (New York: Oxford University Press, 2006).

17. D. Lehmkuhl, *The TIRR Symptom Checklist* (Houston: Institute for Rehabilitation Research, 1998).

18. Medical College of Virginia, *TBI Symptom Checklist* (Richmond: Rehabilitation and Neuropsychological Service, n.d.).

19. M. Picard, D. Scarisbrick and R. Paluck, *HELPS* (New York: Comprehensive Regional TBI Rehabilitation Center, 1991).

20. Mild Traumatic Brain Injury Committee of the Head Injury Interdisciplinary Special Interest Group of the American Congress of Rehabilitation Medicine, "Definition of Mild Brain Injury," *Journal of Head Trauma Rehabilitation* 8 (1993): 86–87.

21. W. A. Gordon, L. Haddad, M. Brown, M. R. Hibbard and M. Sliwinski, "The Sensitivity and Specificity of Self-Reported Symptoms in Individuals with Traumatic Brain Injury," *Brain Injury* 14 (2000): 21–33.

8. THE MEANING OF PSYCHOLOGICAL ABNORMALITY

1. J. Kaufman, B. Z. Yang, H. Douglas-Palumberi, S. Houssyar, D. Lipschitz, J. H. Krystal and J. Gelertner, "Social Supports and Serotonin Transporter Gene Modulate Depression in Maltreated Children," *Proceedings of the National Academy of Sciences* 101 (2004): 17316–17321.

2. M. Rutter, "Incidence of Autism Spectrum Disorders: Changes over Time and Their Meaning," *Acta Paediatrica* 94 (2005): 2–15.

3. J. Kagan, N. Snidman, V. Kahn and S. Towsley, "The Preservation of Two Infant Temperaments into Adolescence," *Monographs of the Society for Research in Child Development* 72 (2007): 1–93.

4. T. N. Crawford, P. Cohen, M. B. First, A. E. Skodol, J. G. Johnson and S. S. Kasen, "Comorbid Axis I and Axis II Disorders in Early Adolescence," *Archives of General Psychiatry* 65 (2008): 641–648.

5. V. Lorant, C. Croux, S. Weich, D. Deliege, J. Mackenbach and M. Ansseau, "Depression and Socio-economic Risk Factors," *British Journal of Psychiatry* 190 (2007): 293–298.

9. THE IMPACT OF MODERN NEUROSCIENCE ON TREATMENT OF PAROLEES: ETHICAL CONSIDERATIONS IN USING PHARMACOLOGY TO PREVENT ADDICTION RELAPSE

1. National Research Council, *Informing America's Policy on Illegal Drugs: What We Don't Know Keeps Hurting Us.* Committee on Data and Research for Policy on Illegal Drugs. Charles F. Manski, John V. Pepper and Carol V. Petrie, eds. Committee on Law and Justice and Committee on National Statistics. Commission on Behavioral and Social Sciences and Education (Washington, DC: National Academy Press, 2001).

2. C. W. Huddleston III, D. B. Marlowe and R. Casebolt, *Painting the Current Picture: A National Report Card on Drug Courts and Other Problem-Solving Court Programs in the United States.* National Drug Court Institute Report, vol. 2, no. 1 (May 2008) (available at http://www.ncjrs.gov/App/Publications/abstract.aspx?ID=245618).

3. MacArthur Foundation Law and Neuroscience Project, http://www.lawandneuroscienceproject.org/.

4. J. W. Cornish, D. Metzger, G. E. Woody, D. Wilson, A. T. McLellan, B. Vandergrift and C. P. O'Brien, "Naltrexone Pharmacotherapy for Opioid Dependent Federal Probationers," *Journal of Substance Abuse Treatment* 14 (1997): 529–534.

5. R. J. Bonnie, "Judicially Mandated Treatment with Naltrexone by Opiate-Addicted Criminal Offenders," *Virginia Journal of Social Policy and the Law* 13 (2005): 64–88.

10. WORKING LATER IN LIFE
MAY FACILITATE NEURAL HEALTH

1. R. S. Wilson and D. A. Bennett. "Cognitive Activity and Risk of Alzheimer's Disease," *Current Directions in Psychological Science* 12 (2003): 87.

2. C. Schooler, M. S. Mulatu and G. Oates. "The Continuing Effects of Substantively Complex Work on the Intellectual Functioning of Older Workers," *Psychology and Aging* 14 (1999): 483–506.

3. D. C. Park and P. Reuter-Lorenz. "The Adaptive Brain: Aging and Neurocognitive Scaffolding," *Annual Review of Psychology* (forthcoming).

11. MANAGING CONFLICTING INTERESTS
IN MEDICAL JOURNAL PUBLISHING

1. J. P. Vandenbroucke, "Medical Journals and the Shaping of Medical Knowledge," *Lancet* 352 (1998): 2001–2006.

2. J. Toy, "The Ingelfinger Rule: Franz Ingelfinger at the NEJM 1967–77," *Science Editor* 25 (2002): 195–198.

3. L. K. Altman, "The Ingelfinger Rule, Embargoes, and Journal Peer Review—Part 1," *Lancet* 347 (1996): 1382–1386.

4. L. K. Altman, "The Ingelfinger Rule, Embargoes, and Journal Peer Review—Part 2," *Lancet* 347 (1996): 1459–1463.

5. M. Errami and H. Garner, "A Tale of Two Citations," *Nature* 451 (2008): 397–399.

6. J. P. Ioannidis, "Why Most Published Research Findings Are False," *PLoS Med* 2005; doi:10.1371/journal.pmed.0020124.

7. D. P. Phillips, E. J. Kanter, B. Bednarczyk et al., "Importance of the Lay Press in the Transmission of Medical Knowledge to the Scientific Community," *New England Journal of Medicine* 325 (1991): 1180–1183.

8. A. Cassels, M. A. Hughes, C. Cole, B. Mintzes, J. Lexchin and J. P. McCormack, "Drugs in the News: An Analysis of Canadian Newspaper Coverage of New Prescription Drugs," *Canadian Medical Association Journal* 168 (2003): 1133–1137.

9. S. L. Hauser and S. C. Johnston, "Of Ghosts and Sirens: The Subtlest Lures of Industry," *Annals of Neurology* 61 (2007): A11–A12.

10. D. de Solla, *Little Science, Big Science ... and Beyond* (New York: Columbia University Press, 1963).

12. PEDIATRIC SCREENING AND THE PUBLIC GOOD

1. N. Forst, "Ethical Implications of Screening Asymptomatic Individuals," *FASEB Journal* 6 (1992): 2813–2817.

2. See ACMG, "Newborn Screening: Toward a Uniform Screening Panel and System," at http://mchb.hrsa.gov/screening/summary.htm.

3. S. M. Myers, C. P. Johnson and the Council on Children with Disabilities, "Management of Children with Autism Spectrum Disorders," *Pediatrics* 120 (2007): 1162–1163.

4. V. A. Moyer et al., "Expanding Newborn Screening: Process, Policy, and Priorities," *Hastings Center Report* 38, no. 3 (2008): 32–39.

5. M. A. Bailey and T. H. Murray, "Ethics, Evidence, and Cost in Newborn Screening," *Hastings Center Report* 38, no. 3 (2008): 23–31.

16. MEMOIRS ABOUT MEMORY: TOO MUCH VERSUS TOO LITTLE

1. E. S. Parker, L. Cahill and J. L. McGaugh, "A Case of Unusual Autobiographical Remembering," *Neurocase* 12 (2006): 35–49.

2. S. Corkin, "What's New with the Amnesic Patient H.M.?" *Nature Reviews Neuroscience* 3 (2002): 153–160.

Index

Other Dana Press Books

www.dana.org/news/danapressbooks

Books for General Readers

Brain and Mind

DEEP BRAIN STIMULATION:
A New Treatment Shows Promise in the Most Difficult Cases
Jamie Talan
An award-winning science writer has penned the first general-audience book to explore the benefits and risks of this cutting-edge technology, which is producing promising results for a wide range of brain disorders.
Cloth• 200 pp • ISBN-13: 978-1-932594-37-9 • $25.00

TRY TO REMEMBER: Psychiatry's Clash Over Meaning, Memory, and Mind
Paul R. McHugh, M.D.
Prominent psychiatrist and author Paul McHugh chronicles his battle to put right what has gone wrong in psychiatry. McHugh takes on such controversial subjects as "recovered memories," multiple personalities, and the overdiagnosis of PTSD.
Cloth • 300 pp • ISBN-13: 978-1-932594-39-3 • $25.00

CEREBRUM 2008: Emerging Ideas in Brain Science
Foreword by Carl Zimmer
The second annual anthology drawn from Cerebrum's highly regarded Web edition, Cerebrum 2008 brings together an international roster of scientists and other scholars to interpret the latest discoveries about the human brain and confront their implications.
Paper •225 pp • ISBN-13: 978-1-932594-33-1 • $14.95

CEREBRUM 2007: Emerging Ideas in Brain Science
Foreword by Bruce S. McEwen, Ph.D.
Paper • 243 pp • ISBN-13: 978-1-932594-24-9 • $14.95

Visit Cerebrum online at www.dana.org/news/cerebrum.

YOUR BRAIN ON CUBS: Inside the Heads of Players and Fans
Dan Gordon, Editor
Our brains light up with the rush that accompanies a come-from-behind win—and the crush of a disappointing loss. Brain research also offers new insight into how players become experts. Neuroscientists and science writers explore these topics and more in this intriguing look at talent and triumph on the field and our devotion in the stands.
6 illustrations.
Cloth • 150 pp • ISBN-13: 978-1-932594-28-7 • $19.95

THE NEUROSCIENCE OF FAIR PLAY:
Why We (Usually) Follow the Golden Rule

Donald W. Pfaff, Ph.D.

A distinguished neuroscientist presents a rock-solid hypothesis of why humans across time and geography have such similar notions of good and bad, right and wrong.
10 illustrations.

Cloth • 234 pp • ISBN-13: 978-1-932594-27-0 • $20.95

BEST OF THE BRAIN FROM SCIENTIFIC AMERICAN:
Mind, Matter, and Tomorrow's Brain

Floyd E. Bloom, M.D., Editor

Top neuroscientist Floyd E. Bloom has selected the most fascinating brain-related articles from *Scientific American* and *Scientific American Mind* since 1999 in this collection.
30 illustrations.

Cloth • 300 pp • ISBN-13: 978-1-932594-22-5 • $25.00

MIND WARS: Brain Research and National Defense

Jonathan D. Moreno, Ph.D.

A leading ethicist examines national security agencies' work on defense applications of brain science, and the ethical issues to consider.

Cloth • 210 pp • ISBN-10: 1-932594-16-7 • $23.95

THE DANA GUIDE TO BRAIN HEALTH:
A Practical Family Reference from Medical Experts (with CD-ROM)

Floyd E. Bloom, M.D., M. Flint Beal, M.D., and David J. Kupfer, M.D., Editors
Foreword by William Safire

A complete, authoritative, family-friendly guide to the brain's development, health, and disorders.
16 full-color pages and more than 200 black-and-white drawings.

Paper (with CD-ROM) • 733 pp • ISBN-10: 1-932594-10-8 • $25.00

THE CREATING BRAIN: The Neuroscience of Genius

Nancy C. Andreasen, M.D., Ph.D.

A noted psychiatrist and best-selling author explores how the brain achieves creative breakthroughs, including questions such as how creative people are different and the difference between genius and intelligence.
33 illustrations/photos.

Cloth • 197 pp • ISBN-10: 1-932594-07-8 • $23.95

THE ETHICAL BRAIN

Michael S. Gazzaniga, Ph.D.

Explores how the lessons of neuroscience help resolve today's ethical dilemmas, ranging from when life begins to free will and criminal responsibility.

Cloth • 201 pp • ISBN-10: 1-932594-01-9 • $25.00

A GOOD START IN LIFE:
Understanding Your Child's Brain and Behavior from Birth to Age 6

Norbert Herschkowitz, M.D., and Elinore Chapman Herschkowitz

The authors show how brain development shapes a child's personality and behavior, discussing appropriate rule-setting, the child's moral sense, temperament, language, playing, aggression, impulse control, and empathy.

13 illustrations.

Cloth • 283 pp • ISBN-10: 0-309-07639-0 • $22.95
Paper (Updated with new material) • 312 pp • ISBN-10: 0-9723830-5-0 • $13.95

BACK FROM THE BRINK:
How Crises Spur Doctors to New Discoveries about the Brain

Edward J. Sylvester

In two academic medical centers, Columbia's New York Presbyterian and Johns Hopkins Medical Institutions, a new breed of doctor, the neurointensivist, saves patients with life-threatening brain injuries.

16 illustrations/photos.

Cloth • 296 pp • ISBN-10: 0-9723830-4-2 • $25.00

THE BARD ON THE BRAIN:
Understanding the Mind Through the Art of Shakespeare and the Science of Brain Imaging

Paul M. Matthews, M.D., and Jeffrey McQuain, Ph.D. • Foreword by Diane Ackerman

Explores the beauty and mystery of the human mind and the workings of the brain, following the path the Bard pointed out in 35 of the most famous speeches from his plays.

100 illustrations.

Cloth • 248 pp • ISBN-10: 0-9723830-2-6 • $35.00

STRIKING BACK AT STROKE: A Doctor-Patient Journal
Cleo Hutton and Louis R. Caplan, M.D.

A personal account, with medical guidance from a leading neurologist, for anyone enduring the changes that a stroke can bring to a life, a family, and a sense of self.

15 illustrations.

Cloth • 240 pp • ISBN-10: 0-9723830-1-8 • $27.00

UNDERSTANDING DEPRESSION:
What We Know and What You Can Do About It

J. Raymond DePaulo, Jr., M.D., and Leslie Alan Horvitz

Foreword by Kay Redfield Jamison, Ph.D.

What depression is, who gets it and why, what happens in the brain, troubles that come with the illness, and the treatments that work.

Cloth • 304 pp • ISBN-10: 0-471-39552-8 • $24.95
Paper • 296 pp • ISBN-10: 0-471-43030-7 • $14.95

KEEP YOUR BRAIN YOUNG:
The Complete Guide to Physical and Emotional Health and Longevity

Guy M. McKhann, M.D., and Marilyn Albert, Ph.D.

Every aspect of aging and the brain: changes in memory, nutrition, mood, sleep, and sex, as well as the later problems in alcohol use, vision, hearing, movement, and balance.

Cloth • 304 pp • ISBN-10: 0-471-40792-5 • $24.95

Paper • 304 pp • ISBN-10: 0-471-43028-5 • $15.95

THE END OF STRESS AS WE KNOW IT

Bruce S. McEwen, Ph.D., with Elizabeth Norton Lasley • Foreword by Robert Sapolsky

How brain and body work under stress and how it is possible to avoid its debilitating effects.

Cloth • 239 pp • ISBN-10: 0-309-07640-4 • $27.95

Paper • 262 pp • ISBN-10: 0-309-09121-7 • $19.95

IN SEARCH OF THE LOST CORD:
Solving the Mystery of Spinal Cord Regeneration

Luba Vikhanski

The story of the scientists and science involved in the international scientific race to find ways to repair the damaged spinal cord and restore movement.

21 photos; 12 illustrations.

Cloth • 269 pp • ISBN-10: 0-309-07437-1 • $27.95

THE SECRET LIFE OF THE BRAIN

Richard Restak, M.D. • Foreword by David Grubin

Companion book to the PBS series of the same name, exploring recent discoveries about the brain from infancy through old age.

Cloth • 201 pp • ISBN-10: 0-309-07435-5 • $35.00

THE LONGEVITY STRATEGY:
How to Live to 100 Using the Brain-Body Connection

David Mahoney and Richard Restak, M.D. • Foreword by William Safire

Advice on the brain and aging well.

Cloth • 250 pp • ISBN-10: 0-471-24867-3 • $22.95

Paper • 272 pp • ISBN-10: 0-471-32794-8 • $14.95

STATES OF MIND: New Discoveries About How Our Brains Make Us Who We Are

Roberta Conlan, Editor

Adapted from the Dana/Smithsonian Associates lecture series by eight of the country's top brain scientists, including the 2000 Nobel laureate in medicine, Eric Kandel.

Cloth • 214 pp • ISBN-10: 0-471-29963-4 • $24.95

Paper • 224 pp • ISBN-10: 0-471-39973-6 • $18.95

The Dana Foundation Series on Neuroethics

DEFINING RIGHT AND WRONG IN BRAIN SCIENCE:
Essential Readings in Neuroethics

Walter Glannon, Ph.D., Editor

The fifth volume in The Dana Foundation Series on Neuroethics, this collection marks the five-year anniversary of the first meeting in the field of neuroethics, providing readers with the seminal writings on the past, present, and future ethical issues facing neuroscience and society.

Cloth • 350 pp • ISBN-10: 978-1-932594-25-6 • $15.95

HARD SCIENCE, HARD CHOICES:
Facts, Ethics, and Policies Guiding Brain Science Today

Sandra J. Ackerman, Editor

Top scholars and scientists discuss new and complex medical and social ethics brought about by advances in neuroscience. Based on an invitational meeting co-sponsored by the Library of Congress, the National Institutes of Health, the Columbia University Center for Bioethics, and the Dana Foundation.

Paper • 152 pp • ISBN-10: 1-932594-02-7 • $12.95

NEUROSCIENCE AND THE LAW: Brain, Mind, and the Scales of Justice

Brent Garland, Editor. With commissioned papers by Michael S. Gazzaniga, Ph.D., and Megan S. Steven; Laurence R. Tancredi, M.D., J.D.; Henry T. Greely, J.D.; and Stephen J. Morse, J.D., Ph.D.

How discoveries in neuroscience influence criminal and civil justice, based on an invitational meeting of 26 top neuroscientists, legal scholars, attorneys, and state and federal judges convened by the Dana Foundation and the American Association for the Advancement of Science.

Paper • 226 pp • ISBN-10: 1-032594-04-3 • $8.95

BEYOND THERAPY: Biotechnology and the Pursuit of Happiness
A Report of the President's Council on Bioethics

Special Foreword by Leon R. Kass, M.D., Chairman

Introduction by William Safire

Can biotechnology satisfy human desires for better children, superior performance, ageless bodies, and happy souls? This report says these possibilities present us with profound ethical challenges and choices. Includes dissenting commentary by scientist members of the Council.

Paper • 376 pp • ISBN-10: 1-932594-05-1 • $10.95

NEUROETHICS: Mapping the Field. Conference Proceedings

Steven J. Marcus, Editor

Proceedings of the landmark 2002 conference organized by Stanford University and the University of California, San Francisco, and sponsored by the Dana Foundation, at which more than 150 neuroscientists, bioethicists, psychiatrists and psychologists, philosophers, and professors of law and public policy debated the ethical implications of neuroscience research findings.

50 illustrations.

Paper • 367 pp • ISBN-10: 0-9723830-0-X • $10.95

Immunology

RESISTANCE: The Human Struggle Against Infection

Norbert Gualde, M.D., translated by Steven Rendall

Traces the histories of epidemics and the emergence or re-emergence of diseases, illustrating how new global strategies and research of the body's own weapons of immunity can work together to fight tomorrow's inevitable infectious outbreaks.

Cloth • 219 pp • ISBN-10: 1-932594-00-0 • $25.00

FATAL SEQUENCE: The Killer Within

Kevin J. Tracey, M.D.

An easily understood account of the spiral of sepsis, a sometimes fatal crisis that most often affects patients fighting off nonfatal illnesses or injury. Tracey puts the scientific and medical story of sepsis in the context of his battle to save a burned baby, a sensitive telling of cutting-edge science.

Cloth • 231 pp • ISBN-10: 1-932594-06-X • $23.95
Paper • 231 pp • ISBN-10: 1-932594-09-4 • $12.95

Arts Education

A WELL-TEMPERED MIND: Using Music to Help Children Listen and Learn

Peter Perret and Janet Fox • Foreword by Maya Angelou

Five musicians enter elementary school classrooms, helping children learn about music and contributing to both higher enthusiasm and improved academic performance. This charming story gives us a taste of things to come in one of the newest areas of brain research: the effect of music on the brain. 12 illustrations.

Cloth • 225 pp • ISBN-10: 1-932594-03-5 • $22.95
Paper • 225 pp • ISBN-10: 1-932594-08-6 • $12.00

Dana Press also offers several free periodicals dealing with brain science, arts education, and immunology. For more information, please visit www.dana.org.